GRAVITATION:
SL (2,C) GAUGE THEORY AND CONSERVATION LAWS

GRAVITATION:
SL (2,C) GAUGE THEORY AND CONSERVATION LAWS

M. Carmeli, E. Leibowitz, and N. Nissani

Ben-Gurion University, Beer-Sheva, Israel

Singapore • New Jersey • London • Hong Kong

Published by
World Scientific Publishing Co. Pte. Ltd.
P O Box 128, Farrer Road, Singapore 9128
USA office: 687 Hartwell Street, Teaneck, NJ 07666
UK office: 73 Lynton Mead, Totteridge, London N20 8DH

GRAVITATION: SL(2,C) GAUGE THEORY AND CONSERVATION LAWS

Copyright © 1990 by World Scientific Publishing Co. Pte. Ltd.

All rights reserved. This book, or parts thereof, may not be reproduced in any form or by any means, electronic or mechanical, including photocopying, recording or any information storage and retrieval system now known or to be invented, without written permission from the Publisher.

ISBN 981-02-0160-5

Printed in Singapore by Utopia Press.

PREFACE

The study of gravitation is one of the most important topics in theoretical physics, because of its implications both on particle and field theories as well as on astrophysics. Of special importance is the formulation of Einstein's general relativity theory as a gauge theory, whereby gravitation is cast as other fields all of which, as is well-known, are described nowadays as gauge fields. The conservation laws in general relativity, from the viewpoint of both Einstein's original classical theory and the $SL(2,C)$ gauge theory of gravitation, is also of great importance since in addition of playing a central role in these theories they bear on the rest of physics.

This monograph deals with the two topics mentioned above; it gives a comprehensive presentation of the $SL(2,C)$ gauge theory of gravitation along with the problem of conservation laws in general relativity. Emphasize is put on quadratic Lagrangians which yield the Einstein field equations, as compared with Hilbert's original linear Lagrangian, thus gravitation follows the other gauge fields all of which are derived from nonlinear Lagrangians. The implications of the quadratic Lagrangians on the conservation laws are then studied. Conserved gravitational currents, which emerge from the $SL(2,C)$ gauge theory of gravitation, are discussed in detail. Noether's theorem is applied to Carmeli and Kaye's spinorial version of Hilbert's Lagrangian with respect to the SL(2,C) group. Nissani and Leibowitz's work on the existence of a preferred class of coordinate systems in which the energy-momentum tensor satisfies a global continuity equation is outlined, and the physical implication of this class of frames is analyzed.

Beer-Sheva, Israel
October, 1989

Moshe Carmeli
Elhanan Leibowitz
Noah Nissani

ACKNOWLEDGEMENTS

We wish to thank our colleagues for the encouragement, helpful remarks and discussions we had during the writing of this monograph. In particular we want to thank Professors N. Rosen, S. Malin, A. Gersten, A. Davidson, J. Bekenstein, Y. Avishai and J. Daboul, and Drs. A. Lonke, A. Guendelman, A. Feinstein and M. Schiffer. We are indebted to the late Mrs. Sara Corrogosky for her assistance, and are particularly grateful to Ms. Y. M. Chong, Editor of World Scientific Publishing Company, for the excellent cooperation we had during the course of printing the book. Last, but not least, we wish to express our thanks to our spouses and families for the unlimited personal help, encouragement, patience and support we received during the writing and publishing of this monograph and without which it would never have been published.

TABLE OF CONTENTS

Preface	v
List of Symbols	xi
1. Introduction	1
2. Fundamentals of the SL(2,C) Gauge Theory of Gravitation	3
2.1 Preliminaries	3
2.2 Spinor space	4
2.2.1 Spinor algebra	4
2.2.2 Spinor analysis	7
2.2.3 Relations between spinors and tensors	9
2.3 Gravitation as an SL(2,C) gauge theory	11
2.3.1 The gauge potential	11
2.3.2 The gauge field	13
2.3.3 The Riemann Tensor	16
2.4 Field equations	19
2.4.1 Free-field equations	19
2.4.2 Coupling with matter	20
2.5 Relation to the Newman-Penrose formalism	21
2.6 Derivation by Palatini's method	25

3. Quadratic Lagrangians 29
   3.1 Linear and quadratic Lagrangians 29
   3.2 Hilbert's Lagrangian 30
   3.3 The quadratic invariants 31
   3.4 Carmeli's Lagrangian 33
   3.5 The Lagrangian family 34
   3.6 The tensorial quadratic Lagrangian 35
      3.6.1 General properties 36
      3.6.2 The variation of $\mathscr{L}$ 39
      3.6.3 The variation of $C_{\alpha\beta\gamma\delta}$ and $T_{\alpha\beta}$ 40
      3.6.4 The variation of the metric tensor 41
   3.7 Summary 42
4. Energy Conservation Problem in General Relativity 45
   4.1 Introduction 45
   4.2 The problem 46
   4.3 Einstein's solution 49
   4.4 Landau-Lifshitz pseudo-tensor 53
   4.5 Moller's approach 56
   4.6 Conserved tensor density — gauge formulation approach 60
   4.7 Conclusions 62
5. SL(2,C) Conservation Laws of Gravitation 63
   5.1 Symmetry and Conservation Laws 63
   5.2 The set of six currents 64
   5.3 The SL(2,C) symmetry 66
   5.4 The gravitational current 68
   5.5 Example: the Tolman metric 71
   5.6 Summary 75
6. Non-rotating Frames 77
   6.1 Preliminaries 77
   6.2 The nonrotating coordinate systems 78
   6.3 The internal group 81
   6.4 Inertial coordinates 83
   6.5 Example 86
   6.6 Conclusions and remarks 88
References 91
Subject Index 95

LIST OF SYMBOLS

Symbol	*Description*
S	Element of the group SL(2,C)
ϕ^A	2-component spinor
ε_{AB}	Levi-Civita metric
$\xi_a{}^A$	Spin frame
$\Gamma_{A\mu}{}^B$	Spinor affine connection
$\sigma^\mu_{AB'}$	Hermitian matrices
$\varepsilon^{\alpha\beta\gamma\delta}$	Levi-Civita skew-symmetric tensor density
$\alpha, \ldots, \gamma$	Spin coefficients
Ψ_{ABCD}	Weyl spinor
$\phi_{ABC'D'}$	Trace-free Ricci spinor
$\Psi_0 \ldots \Psi_4$	Components of Weyl spinor
$\phi_{00} \ldots \phi_{22}$	Components of trace-free Ricci spinor
Λ	Ricci's scalar/24
$T^{\alpha\beta}$	Energy-momentum tensor
$\Gamma^\gamma_{\alpha\beta}$	Christoffel symbols

Symbol	Description
$G^{\alpha\beta}$	Einstein's tensor
$R^{\alpha\beta}$	Ricci's tensor
R	Ricci's scalar
$g_{\alpha\beta}$	The metric tensor
κ	Einstein's gravitational constant
g	Determinant of the metric tensor
$\tau^{\mu\nu}$	Gravitational energy-momentum pseudo-tensor
$\mathscr{L}$	Lagrangian
$\mathscr{L}_E$	Einstein's Lagrangian
$\mathscr{L}_H$	Hilbert's Lagrangian
$\tau_E{}^\mu{}_\nu$	Einstein's gravitational energy-momentum pseudo-tensor
$\boldsymbol{\tau}_E{}^\mu{}_\nu$	Einstein's total energy-momentum complex
$\mathscr{L}_M$	Matter Lagrangian
P^α	Energy-momentum vector
$\boldsymbol{\tau}_L{}^\mu{}_\nu$	Landau-Lifshitz's total energy-momentum complex
$\tau_L{}^\mu{}_\nu$	Landau-Lifshitz's gravitational energy-momentum pseudo-tensor
$M^{\alpha\beta\gamma}$	Angular-momentum tensor
$\Phi_L{}^{\mu\nu\rho}$	Landau-Lifshitz's super-potential
$\Phi_E{}^{\mu\nu\rho}$	Freud's super-potential
$\Phi_M{}^{\mu\nu\rho}$	Moller's super-potential
$\tau_M{}^\nu{}_\mu$	Moller's total energy-momentum pseudo-tensor
$\boldsymbol{\tau}_M{}^\nu{}_\mu$	Moller's gravitational energy-momentum complex
$h^A{}_\mu$	Moller's tetrad
$\theta_{\mu\nu\rho}$	Moller's tensor
ϕ^ρ	Moller's vector
$\bar{\tau}^\nu_{M\mu}$	Moller's gravitational energy-momentum second pseudo-tensor
$\bar{\Phi}_M{}^{\mu\nu\rho}$	Moller's second super-potential
$F_{\mu\nu a}{}^b$	Carmeli's gauge field tensor
$B_{\mu a}{}^b$	Carmeli's gauge potential vector
$R^\rho{}_{\sigma\mu\nu}$	Riemann's curvature tensor
$\sigma_{\rho ac'}$	The metric tetrad
$L_{\mu\nu}$	Gauge potential rotor tensor
$K_{\mu\nu}$	Gauge potential commutator tensor
$*L^{\alpha\beta}$	Antisymmetric conserved tensor density
$\Phi^{\alpha\beta\mu}$	Super-potential
$\mathscr{L}_0$	Hilbert's Lagrangian in vacuum
$\mathscr{L}_{Li}$	Quadratic scalar densities

List of Symbols

$\mathcal{L}_{Li}$	($i = 1, 2, 3$) High-order theories Lagrangians
$\mathcal{L}_{Li}$	($i = 4, 5$) Lanczos' invariants
$\mathcal{L}_C$	Carmeli's Lagrangian
F^W	Weyl's part of Carmeli's gauge field tensor
F^R	Ricci's part of Carmeli's gauge field tensor
$\mathcal{L}_K$	Spinorial family of Lagrangians
$U_{\alpha\beta\gamma\delta}$	Four-index energy-momentum tensor
$C_{\alpha\beta\gamma\delta}$	Four-index gravitational tensor (Weyl)
$\mathcal{L}_i$	Scalar density components of the tensorial quadratic Lagrangian
$\mathcal{L}_N$	Tensorial quadratic Lagrangian
$\mathcal{L}_F$	Tensorial quadratic Lagrangian in vacuum
P^μ	Energy-momentum affine vector
J_C	Carmeli's conserved vector density
J_M	Malin's conserved vector density
F^S	Part of Carmeli's gauge field obtained from Ricci's traceless tensor
F^T	Part of Carmeli's gauge field obtained from Ricci's scalar
J_σ	The "gravitational" conserved current

The metric signature used is $(1, -1, -1, -1)$

Dios guarda entre sus secretos
el secreto que eso encierra
y mando que todo peso
cayera siempre a la tierra

Jose Hernandez, *Martin Fierro*,
part II, line 4335 (1872).

God keeps among his secrets
the secret that lies here
and ordered that each object
should always fall to the ground

1
INTRODUCTION

Einstein's general relativity is probably the most ingenious theory invented by man so far[1]. Gravitation became one of the most important topics in theoretical physics because of its implications both on particle physics and field theory as well as on astrophysics. Of special importance is the formulation of Einstein's original theory as a gauge theory, whereby gravitation is cast as the other physical fields all of which, as is well-known, are described nowadays as gauge fields. The conservation laws in general relativity, from the viewpoint of both Einstein's original classical theory and the more newly developed SL(2,C) gauge theory of gravitation, is also of great importance since in addition to playing a central role in these theories they bear on the rest of physics.

This monograph deals with the two topics mentioned above; it gives a comprehensive presentation of the SL(2,C) gauge theory of gravitation[2-22] along with the problem of conservation laws in general relativity[23-31]. Emphasize is put on quadratic Lagrangians which yield the Einstein field equations, as compared with Hilbert's original linear Lagrangian[32], thus gravitation follows the other gauge fields all of which are derived from nonlinear Lagrangians. The implications of the quadratic Lagrangians on the conservation laws are then studied. Conserved gravitational currents, which emerge from the SL(2,C) gauge theory of gravitation, are discussed in detail. Noether's theorem is applied to Carmeli and Kaye's spinorial version of Hilbert's Lagrangian with respect to the SL(2,C) group. The existence of a preferred class of coordinate systems, in which the energy-momentum tensor satisfies a global continuity equation, is outlined, and the physical implication of these frames is analyzed.

In Chapter 2, we review the SL(2,C) gauge theory of gravitation[2-22]. This theory is a version of relativistic physics, modeled after the theory of Yang and Mills[33], which brings gravity in line with other physical fields. The main features of the theory will be presented.

A fiber bundle formulation of the SL(2,C) gauge theory of gravitation was given by Kaye[21]. In this theory the spinor affine space-time is taken as the base space of a fiber bundle, where the basic objects are two 2-component spinors at each point of space-time[34]. Einstein's theory of gravitation in the Newman-Penrose formalism[35] is obtained from the SL(2,C) gauge theory in the same spinor space and Riemannian space-time.

In Chapter 3 we review recent developments in the theory of quadratic-Lagrangian formulation of general relativity theory[4,13,15,28,29,36,37]. Of special interest will be the quadratic invariants obtained from the Riemann tensor, and the question whether they are suitable to serve as Lagrangians.

In Chapter 4, a review is given to the problems associated with the conservation laws in general relativity theory. A handful of the many suggestions on how to tackle the problem, proposed in the past seventy years, will be outlined. The conserved skew-symmetric tensor density will be discussed as well.

In Chapter 5, we discuss the conservation laws associated with the SL(2,C) gauge theory of gravitation. We show the existence of a set of six conserved currents[24-26]. These includes the currents of Carmeli[13] and that of Malin[38], and several of them are obtainable by applying Noether's theorem[39] to Carmeli's and to Carmeli-Kaye's Lagrangians. As an example we calculate the gravitational current for the special case of Tolman's space[40].

Preferred coordinate systems in curved space-time in which the energy-momentum tensor is globally conserved are finally introduced, following Nissani and Leibowitz[30,31], in Chapter 6. The study of these coordinate systems could be important in connection with the conservation laws in general relativity and is also of great interest *per se*.

2
FUNDAMENTALS OF THE SL (2,C) GAUGE THEORY OF GRAVITATION

2.1 *Preliminaries*

At the heart of general relativity lies the principle of general covariance, which attests to the equivalence of all coordinate systems. This principle postulates that the laws of physics are invariant under the group of general coordinate transformations (subject to appropriate differentiability conditions), called the manifold mapping group (MMG)[41]. Since tensors transform under representations of the MMG, they are natural objects to serve as the building blocks of a covariant physical theory. Considerations of the covariance requirements lead to the definition of the covariant derivative in terms of affine connections (the Christoffel symbols). When the latter are interpreted as potentials, they give rise, through the standard procedures of non-Abelian gauge theory, to the Riemann curvature tensor.

$$R^\delta_{\alpha\beta\gamma} = \Gamma^\delta_{\alpha\gamma,\beta} - \Gamma^\delta_{\alpha\beta,\gamma} + \Gamma^\mu_{\alpha\gamma}\Gamma^\delta_{\mu\beta} - \Gamma^\mu_{\alpha\beta}\Gamma^\delta_{\mu\gamma} , \qquad (2.1)$$

which is the basic tensor of the relativistic theory of gravitation. On both mathematical and physical grounds, the thrust of the theory of gravity is the study of this tensor. It turns out, therefore, that Einstein's theory of gravitation, in its classical formulation, is largely an MMG gauge theory, even if historically it had not been conceived as such.

On the other hand, general relativity is characterized by another postulate—the principle of equivalence, which asserts that the laws of special relativity remain, locally, valid in curved space-time of general relativity. At each point of the space-time manifold, according to this principle, there exists a special coordinate system in which the equations expressing the physical laws retain their special relativistic form—the geodesic system of coordinates.

Special relativity embodies the invariance of the laws of nature under Lorentz transformations. Accordingly, in view of the fact that the group SL(2,C) of unimodular transformations of a two-dimensional complex vector space is a double-valued representation of the proper (homogeneous) Lorentz group, it appears natural to attempt to formulate general relativity in terms of the group SL(2,C). This undertaking is accomplished by the SL(2,C) gauge theory of gravitation due to Carmeli[2-22]. This version of relativistic physics, modeled after the theory of Yang and Mills[33], and which brings gravity in line with the other physical fields, is the concern of the present chapter. The main features of the theory will be introduced.

In the next section spinor space is introduced, on which spinor algebra and spinor analysis are developed, and the correspondence between spinors and tensors is specified. The basic constituents of the SL(2,C) gauge theory of gravitation—the gauge potential and gauge field—are enunciated in Sec. 2.3, and they are tied up with the curvature of the underlying manifold. In Sec. 2.4 the gravitational field equations are derived from a variational principle with a nonlinear Lagrangian, which subsequently in Sec. 2.5 are shown to yield the Newman-Penrose equations. Finally, Sec. 2.6 is devoted to using Palatini's method to derive the gravitational field equations within the framework of the SL(2,C) gauge theory.

2.2 Spinor Space

2.2.1 Spinor algebra

As explained above, the existence of a two-to-one homomorphism between the time-preserving, reflection-free Lorentz group, and the group SL(2,C) of unimodular linear transformations, gave impetus to the development of spinor calculus in flat 4-dimensional manifold, which was subsequently generalized to curved space-time[42].

At each point of the 4-dimensional space-time, a complex 2-dimensional linear space, called the spinor space, is defined. The elements of the spinor space are 2-dimensional spinors, namely, 2-component complex quantities ϕ^A; $A = 0, 1$ (Latin indices throughout this chapter assume the values 0 and 1). The elements of the complex conjugate spinor space will be labeled with primed Latin upper-case indices ($\phi^{A'}$). In order to get representations of the orthochronous Lorentz group, the 2-component spinors are subject to the group of unimodular spin transformations, namely, the group SL(2,C) of two-by-two complex matrices

$$S = \begin{pmatrix} \alpha & \beta \\ \gamma & \delta \end{pmatrix}, \tag{2.2}$$

with unit determinant,

$$|S| = \alpha\delta - \gamma\beta = 1. \tag{2.3}$$

Thus, if ϕ^A is a spinor, then it undergoes the spinor transformation

$$\phi'^A = \phi^B S_B{}^A. \tag{2.4}$$

The elements of the complex conjugate spinor space are subject to the transformation

$$\phi'^{A'} = \phi^{B'} \overline{S}_{B'}{}^{A'},$$

determined by the complex conjugate matrices

$$\overline{S}_{B'}{}^{A'} = \overline{S_B{}^A}. \tag{2.5}$$

Spinors of higher valence transform according to the tensor product representation, e.g.,

$$\phi'^{AB} = \phi^{CD} S_C{}^A S_D{}^B.$$

In analogy with tensor calculus, in addition to the contravariant spinors ϕ^A and $\phi^{A'}$, another kind of spinors, called covariant, are introduced, and they are labeled by subscripts. The role of the metric tensor, for the purpose of raising and lowering indices, is played here by the antisymmetric matrix

$$\varepsilon_{AB} = \varepsilon^{AB} = \varepsilon_{A'B'} = \varepsilon^{A'B'} = \begin{pmatrix} 0 & 1 \\ -1 & 0 \end{pmatrix}, \qquad (2.6)$$

which is the 2-dimensional alternating (Levi-Civita) symbol, i.e.,

$$\varepsilon_{AB} = -\varepsilon_{BA}, \qquad \varepsilon_{01} = 1,$$

with similar expressions for $\varepsilon_{A'B'}$, ε^{AB} and $\varepsilon^{A'B'}$. These matrices satisfy the relations

$$\varepsilon_{AC}\varepsilon^{BC} = \delta_A^B. \qquad (2.7)$$

Notice that by the definition of a determinant,

$$\varepsilon^{CD}|S| = \varepsilon^{AB} S_A{}^C S_B{}^D,$$

and in view of the unimodularity of S, this implies

$$\varepsilon^{CD} = \varepsilon^{AB} S_A{}^C S_B{}^D. \qquad (2.8)$$

Hence the matrix ε^{AB}, if interpreted as a contravariant spinor of valence 2, is invariant under spinor transformations, which is a necessary condition for the consistency of (2.6). Now spinor indices are lowered and raised according to the rules

$$\phi_A = \phi^B \varepsilon_{BA}, \qquad (2.9)$$

and

$$\phi^A = \varepsilon^{AB} \phi_B, \qquad (2.10)$$

(note the order of the indices). Any one of these two equations implies the other when (2.7) is taken into account. The matrices ε have been written here once as a right multiplier and once as a left multiplier, and under such a convention the summation is always performed over repeated indices.

The transformation law for a covariant spinor ϕ_A is determined by the requirement that scalar products are preserved,

$$\phi_A \phi^A = \phi'_A \phi'^A ,$$

which implies

$$\phi_A = S_A{}^B \phi'_B . \tag{2.11}$$

Thus covariant spinors transform with the inverse matrix S^{-1}. Similarly, the transformation laws for $\phi_{A'}$, or for general mixed spinors such as $\phi_{AB'}{}^C$, can be written down.

A normalized spinor basis is any pair of spinors ("dyad"),

$$\eta_a^A = (\eta_0^A, \eta_1^A) , \tag{2.12}$$

satisfying the normalization condition

$$\eta_a^A \varepsilon_{AB} \eta_b^B = \varepsilon_{ab} = \begin{pmatrix} 0 & 1 \\ -1 & 0 \end{pmatrix} , \tag{2.13}$$

which also implies

$$\eta_a^A \varepsilon^{ab} \eta_b^B = \eta^{AB} , \tag{2.14}$$

with $\varepsilon^{ab} = \varepsilon_{ab}$. Notice that this normalization condition, in view of (2.7), is invariant under the action of the group SL(2,C). An example for a normalized spinor basis is the dyad

$$\eta_0^A = (1, 0) , \quad \eta_1^A = (0, 1) . \tag{2.15}$$

2.2.2 Spinor analysis

So far, the spinor space at one point of the base manifold has been considered, and no relation has been imposed between the spinor representations at two different points of space-time. If we now consider the section of spinors $\phi_A(x)$ over a neighborhood in space-time, the question arises as to how they behave under change of coordinates in the base space. *A priori* it is possible that the spinor be, as in the isotopic spin formalism, scalars under the group of coordinate transformations. Accordingly if $L \in$ MMG, then

$$L(\phi^A(x)) = \phi'^A(x') = \phi^A(x) \,. \tag{2.16}$$

Alternatively, one may assume that the action of L induces a transformation $S(L)$ of the spinor space, whereby

$$L(\phi^A(x)) = \phi^B S(L)_B{}^A \,. \tag{2.17}$$

Obviously, the particular choice

$$S(L)_A{}^B = \delta_A{}^B \tag{2.18}$$

reduces the latter case to the first. In spite of the fact that the group of spinor transformations is the universal covering group of the orthochronous Lorentz group, which is a subgroup of the MMG, there remains a complete freedom to choose between the two options[38]. Thus, no inconsistency will arise if we stick to the convention (2.18), which renders the spinors, as in the Yang-Mills gauge theory, scalars under the action of the MMG.

In order, however, to be able to compare spinors at different points, we resort to the concept of covariant derivatives of spinors. One defines

$$\begin{aligned}\phi_{A;\mu} &= \phi_{A,\mu} - \Gamma_{A\mu}^{B}\phi_B \,, \\ \phi^A{}_{;\mu} &= \phi^A{}_{,\mu} + \Gamma_{B\mu}^{A}\phi^B \,,\end{aligned} \tag{2.19}$$

where $\Gamma_A{}^B{}_\mu$, with $\mu = 0, 1, 2, 3$, are four two-by-two complex matrices, yet to be specified, which are called spinor affine connections, and are supposed to transform like vectors under coordinate transformations. Consequently the covariant derivatives $\phi^A{}_{;\mu}$ and $\phi_{A;\mu}$ behave as vectors under coordinate transformations in space-time, and like spinors under spinor transformations S, provided that the spinor affine connections are subject to the transformation law

$$\Gamma'^B_{A\mu} = (S^{-1})_A{}^D \Gamma_{D\mu}^{C} S_C{}^B + (S^{-1})_A{}^C{}_{,\mu} S_C{}^B \,. \tag{2.20}$$

The formula (2.19), together with the requirement that covariant differentiation obeys the Leibniz rule for product, entail that the covariant derivative of the scalar product of spinors reduces to the ordinary partial derivative. The operation of covariant differentiation generalizes to spinors of higher valence, involving any number of primed and unprimed, covariant and contravariant indices, in an obvious manner.

A natural requirement, which simplifies the manipulations of spinor analysis considerably, is

$$\varepsilon_{AB;\mu} = 0 \, , \tag{2.21}$$

which in view of (2.13) and (2.19), translates to

$$\Gamma_{A\mu}^{\ C}\varepsilon_{CB} + \Gamma_{B\mu}^{\ C}\varepsilon_{AC} = 0 \, . \tag{2.22}$$

In particular,

$$\Gamma_{A\mu}^{\ A} = 0 \, , \tag{2.23}$$

i.e., the spinor affine connections are traceless matrices. There still remains a great deal of freedom in the choice of the spinor affine connections. They are uniquely determined after a further condition is imposed in the next section.

2.2.3 Relations between Spinors and Tensors

Up to this point, the metric structure of space-time and the spinor space co-exist independently. A correspondence between the two is established in terms of connecting mixed quantities $\sigma^\mu{}_{AB'}$. The latter constitute a Hermitian spinor vector, a set of four Hermitian two-by-two matrices

$$\sigma^\mu_{AB'} = \bar{\sigma}^\mu_{B'A} \, ; \quad \mu = 0, 1, 2, 3 \, .$$

These mixed quantities are assumed to transform as space-time vectors in the index μ and as spinors in the indices A, B'. Furthermore, they are subject to the relations

$$\sigma^\mu_{AB'}\sigma_\nu^{AB'} = \delta^\mu_\nu \, , \quad \sigma^\mu_{AB'}\sigma_\mu^{CD'} = \delta^C_A \delta^{D'}_{B'} \, . \tag{2.24}$$

Note that these algebraic conditions imply the "anti-commutation relations"

$$\sigma^\mu_{AC'}\sigma^{\nu BC'} + \sigma^\nu_{AC'}\sigma^{\mu BC'} = g^{\mu\nu}\delta^B_A \, .$$

It is evident that the space-time metric tensor is completely determined once the connecting quantities are given. In flat space-time and Minkowskian coordinates it is possible to choose the connecting quantities (up to a numerical coefficient) as the three Pauli matrices together with the unit matrix, viz.,

$$\sigma^0_{AB'} = \frac{1}{\sqrt{2}}\begin{pmatrix} 1 & 0 \\ 0 & 1 \end{pmatrix}, \qquad \sigma^1_{AB'} = \frac{1}{\sqrt{2}}\begin{pmatrix} 0 & 1 \\ 1 & 0 \end{pmatrix},$$

$$\sigma^2_{AB'} = \frac{1}{\sqrt{2}}\begin{pmatrix} 0 & i \\ -i & 0 \end{pmatrix}, \qquad \sigma^3_{AB'} = \frac{1}{\sqrt{2}}\begin{pmatrix} 1 & 0 \\ 0 & -1 \end{pmatrix}. \qquad (2.25)$$

Moreover, in curved space-time, at an arbitrary point, there exist coordinate systems in the underlying manifold and spinor frames in the spinor space such that at the given point $\sigma^\mu{}_{AB',\alpha} = 0$ and the σ's assume their special relativistic form (2.25).

The mapping between tensors and spinors is accomplished by assigning to a vector V_μ the Hermitian spinor $V_{AB'}$ and vice versa by

$$V_{AB'} = V_\mu \sigma^\mu_{AB'}, \qquad V_\mu = V_{AB'}\sigma_\mu^{AB'}. \qquad (2.26)$$

The generalization to higher-valence tensors is obvious, e.g.,

$$V_{AB'CD'} = V_{\mu\nu}\sigma^\mu_{AB'}\sigma^\nu_{CD'}. \qquad (2.27)$$

It is easily verified that

$$g_{\mu\nu} \leftrightarrow \varepsilon_{AC}\varepsilon_{B'D'}.$$

The connecting quantities $\sigma^\mu{}_{AB'}$ provide a natural way to determine the spin connections $\Gamma_A{}^B{}_\mu$, which up to this point, as remarked in the previous subsection, have remained with a great deal of freedom. One imposes the requirement

$$\sigma^\mu_{AB';\nu} = 0, \qquad (2.28)$$

which together with (2.21) determines the spinor affine connections as functions of the Christoffel symbols, the σ's, and their derivatives:

$$\Gamma_{A\mu}^C = \frac{1}{2}\sigma_\nu^{CB'}(\sigma^\rho_{AB'}\Gamma^\nu_{\rho\mu} + \sigma^\nu_{AB',\mu}). \qquad (2.29)$$

Fundamentals of The SL(2,C) Gauge Theory of Gravitation

As was already shown, given an arbitrary point in space-time, there exist a coordinate system and a spinor frame so that the ordinary derivatives of the connecting quantities $\sigma^{\mu}{}_{AB'}$ vanish at the point. Choosing such a coordinate system and a spinor frame, in view of Eq. (2.24), the derivatives of the metric tensor and hence the affine connections vanish at the point, and Eq. (2.29) shows that the spinor affine connections vanish at the point too.

2.3 Gravitation as an SL(2,C) Gauge Theory

2.3.1 The gauge potential

Let $\eta_a{}^A$ be a normalized spinor basis, i.e., a pair of spinors subject to the normalization condition

$$\varepsilon_{AB}\eta_a^A\eta_b^B = \varepsilon_{ab} \tag{2.30}$$

(a and b are dyad indices). Then any spinor can be expanded in terms of the dyad $\eta_a{}^A$, viz.,

$$\phi^A = \phi^a \eta_a^A . \tag{2.31}$$

The scalars ϕ^a are called the dyad components of the spinor ϕ^A. In particular, the covariant derivatives $\eta_a{}^A{}_{;\mu}$, being mixed quantities, possessing spinorial character with respect to the index A and vectorial character with respect to the index μ, have the representation

$$\eta^A_{a;\mu} = B_{a\mu}^{\ b} \eta_b^A . \tag{2.32}$$

The coefficients $B_{a\ \mu}^{\ b}$, with $\mu = 0, 1, 2, 3$, which form a set of four 2×2 complex matrices, will be taken as the gauge potentials of the theory. Contracting (2.32) with $\eta^a{}_A$ yields

$$B_{a\ \mu}^{\ a} = 0 , \tag{2.33}$$

i.e., the gauge potential matrices are traceless.

Evidently, the gauge potential behaves as a vector with respect to coordinate transformations in space-time. We now examine its behavior under a "gauge transformation", namely, under change of the normalized spinor basis. If $\eta'_a{}^A$ is another normalized spinor basis, it must be related to the $\eta_a{}^A$ basis by

$$\eta_a^A = S_a{}^c \eta_c^{\prime A}, \tag{2.34}$$

where $S_a{}^c$ is an element of the group SL(2,C). Substituting this in (2.32) yields

$$\eta_{a;\mu}^A = (S_a{}^c \eta_c^{\prime A})_{;\mu} = B_{a\mu}^{\ b} \eta_b^A = B_{a\mu}^{\ b} S_b{}^d \eta_d^{\prime A}, \tag{2.35}$$

therefore

$$S_{a,\mu}^{\ c} \eta_c^{\prime A} + S_a{}^c \eta_{c;\mu}^{\prime A} = B_{a\mu}^{\ b} S_b{}^d \eta_d^{\prime A}. \tag{2.36}$$

Hence, one finds

$$\eta_{c;\mu}^{\prime A} = (S^{-1})_c{}^a (B_{a\mu}^{\ b} S_b{}^d - S_{a,\mu}^{\ d}) \eta_d^{\prime A}. \tag{2.37}$$

On the other hand, by (2.32),

$$\eta_{c;\mu}^{\prime A} = B_{c\mu}^{\prime d} \eta_d^{\prime A}.$$

Comparing with (2.37), therefore, we obtain the transformation law for the gauge potential:

$$B_{c\mu}^{\prime d} = (S^{-1})_c{}^a (B_{a\mu}^{\ b} S_b{}^d - S_{a,\mu}^{\ d}). \tag{2.38}$$

This transformation, as is to be expected for gauge potentials, is nonhomogeneous.

The gauge potential is related to the spinor affine connections through (2.32), viz.,

$$\eta_{a;\mu}^A = \eta_{a,\mu}^A + \eta_a^C \Gamma_{C\mu}^{\ A} = B_{a\mu}^{\ b} \eta_b^A. \tag{2.39}$$

In particular, in a normalized spinor basis independent of the coordinates (e.g., (3.15)):

$$B_{a\mu}^{\ c} = \Gamma_{A\mu}^{\ C} \eta_a^A \eta_C^c,$$

i.e., the gauge potential in this case is the set of dyadic components of the spinor affine connections.

2.3.2 *The gauge field*

With the aid of the gauge potential $B_a{}^b{}_\mu$, the gauge field of the theory is now defined by:

$$F_{a\mu\nu}^{\ b} = B_{a\mu,\nu}^{\ b} - B_{a\nu,\mu}^{\ b} + B_{a\mu}^{\ c}B_{c\nu}^{\ b} - B_{a\nu}^{\ c}B_{c\mu}^{\ b}. \tag{2.40a}$$

In the sequel we will often suppress dyad indices. Thus for example, Eq. (2.40a) will be written in the matrix form

$$F_{\mu\nu} = B_{\mu,\nu} - B_{\nu,\mu} + [B_\mu, B_\nu], \tag{2.40b}$$

where

$$[B_\mu, B_\nu] = B_\mu B_\nu - B_\nu B_\mu$$

is the commutator of the two matrices. Note that the partial derivatives in the definition (2.40) of the gauge field $F_{\mu\nu}$ can be replaced by covariant derivatives, viz.,

$$F_{\mu\nu} = B_{\mu;\nu} - B_{\nu;\mu} + [B_\mu, B_\nu], \tag{2.41}$$

thereby exhibiting the tensorial character of the field $F_{\mu\nu}$. As to the behavior of the field under gauge transformation, it follows from (2.38) that

$$F'{}_{a\mu\nu}^{\ b} = (S^{-1})_a{}^c F_{c\mu\nu}^{\ d} S_d{}^b. \tag{2.42}$$

The geometrical meaning of the gauge field will emerge upon examining the commutator of the covariant derivatives. Differentiating (2.32) covariantly and using (2.40) yield:

$$\eta^A_{a;\mu\nu} - \eta^A_{a;\nu\mu} = F_{a\mu\nu}^{\ b}\eta^A_b. \tag{2.43}$$

Let $F_A{}^B{}_{\mu\nu}$ denote the spinor-tensor whose dyad components are (2.40):

$$F_{a\mu\nu}^{\ b} = -F_{A\mu\nu}^{\ B}\eta^A_a \eta^b_B, \tag{2.44}$$

namely,

$$F_{A\mu\nu}^{\ B} = -F_{a\mu\nu}^{\ b}\eta^a_A \eta^B_b. \tag{2.45}$$

Then, for an arbitrary spinor ϕ^A, the commutator of the covariant derivatives satisfies:

$$\phi^A{}_{;\mu\nu} - \phi^A{}_{;\nu\mu} = \phi^B F_{B\mu\nu}^A . \tag{2.46}$$

In deriving (2.46), use has been made of the fact that

$$\phi^a{}_{;\mu\nu} - \phi^a{}_{;\nu\mu} = \phi^a{}_{,\mu\nu} - \phi^a{}_{,\nu\mu} = 0 . \tag{2.47}$$

Since the dyad components ϕ^a are scalars, Eq. (2.46) reveals that $F_A{}^B{}_{\mu\nu}$ is the curvature spinor-tensor. In particular, if $F_A{}^B{}_{\mu\nu}$ is not identically zero then the manifold does not admit a parallel spinor.

Applying the rule (2.46) to the spinor ε_{AB} and taking into account (2.21), we get

$$F_{AB\mu\nu} = F_{BA\mu\nu} , \tag{2.48}$$

and

$$F_A{}^A{}_{\mu\nu} = 0 . \tag{2.49}$$

Thus, the curvature spinor tensor is antisymmetric in the space-time indices μ, ν and symmetric in the spinor indices A, B.

An explicit expression of the curvature spinor-tensor in terms of the spinor affine connections is obtained by observing that differentiating Eq. (2.19) leads to

$$\phi^A{}_{;\mu\nu} - \phi^A{}_{;\nu\mu} = (\Gamma_{B\mu,\nu}^A - \Gamma_{B\nu,\mu}^A + \Gamma_{B\mu}^C \Gamma_{C\nu}^A - \Gamma_{B\nu}^C \Gamma_{C\mu}^A)\phi^B . \tag{2.50}$$

Comparison with (2.46) yields

$$F_{A\mu\nu}^B = \{\Gamma_{\mu,\nu} - \Gamma_{\nu,\mu} + [\Gamma_\mu, \Gamma_\nu]\}_A{}^B . \tag{2.51}$$

Since the gauge field (2.40) has been shown to be associated with the curvature, it is expected to satisfy Bianchi's identities. In order to derive them, we will calculate $\varepsilon^{\alpha\beta\gamma\delta} F_a{}^b{}_{\beta\gamma;\delta}$, where ε is the totally antisymmetric Levi-Civita tensor density whose value is $+1$ (-1) if $\alpha\beta\gamma\delta$ is an even (odd) permutation of 0123, and zero otherwise. First we observe that due to the symmetry of the affine connections, the covariant derivatives in this expression can be replaced by ordinary partial derivatives, and therefore, by (2.40):

$$\varepsilon^{\alpha\beta\gamma\delta} F^{b}_{a\beta\gamma;\delta} = \varepsilon^{\alpha\beta\gamma\delta} F_{\beta\gamma,\delta} = \varepsilon^{\alpha\beta\gamma\delta}(B_{\beta,\gamma\delta} - B_{\gamma,\beta\delta}) + \varepsilon^{\alpha\beta\gamma\delta}[B_\beta, B_\gamma]_{,\delta} \,. \tag{2.52}$$

The first term on the right-hand side vanishes due to the symmetry of the second order derivatives, while in the second term we substitute

$$[B_\beta, B_\gamma]_{,\delta} = [B_{\beta,\delta}, B_\gamma] + [B_\beta, B_{\gamma,\delta}] \,, \tag{2.53}$$

and use the properties of the tensor $\varepsilon^{\alpha\beta\gamma\delta}$, to get

$$\varepsilon^{\alpha\beta\gamma\delta} F_{\beta\gamma;\delta} = \varepsilon^{\alpha\beta\gamma\delta}[B_\delta, (B_{\beta,\gamma} - B_{\gamma,\beta})] \,. \tag{2.54}$$

By (2.40),

$$\varepsilon^{\alpha\beta\gamma\delta}[B_\delta, (B_{\beta,\gamma} - B_{\gamma,\beta})] = \varepsilon^{\alpha\beta\gamma\delta}[B_\delta, F_{\beta\gamma}] \tag{2.55}$$

(since $\varepsilon^{\alpha\beta\gamma\delta}[B_\delta, [B_\beta, B_\gamma]] = 0$). Finally, substituting (2.55) into (2.54) yields the identities

$$\varepsilon^{\alpha\beta\gamma\delta}(F_{\beta\gamma,\delta} + [F_{\beta\gamma}, B_\delta])_a{}^b = 0 \,. \tag{2.56}$$

The similarity of (2.56) to the Bianchi identities of the Riemann curvature tensor becomes more transparent when (2.56) is translated from the dyad components representation to the spinor components language. Indeed, the identities (2.56) assume the form

$$\varepsilon^{\alpha\beta\gamma\delta}(F_{\beta\gamma,\delta} + [F_{\beta\gamma}, \Gamma_\delta])_A{}^B = 0 \,. \tag{2.57}$$

At an arbitrary point O in space-time, one can always choose a coordinate system and a spinor frame such that both the affine connections and the spin connections vanish at the point O. Then, at the point O, Eq. (2.57) is equivalent to

$$\varepsilon^{\alpha\beta\gamma\delta} F_{\beta\gamma;\delta} = 0 \,. \tag{2.58}$$

Since Eq. (2.58) is invariant under coordinate and spinor transformations, it is valid in any coordinate system and any spinor frame. Equation. (2.58) is the spinor version of the Bianchi identities, once the explicit relation between the gauge field and the Riemann curvature tensor is established.

It is also possible to prove (2.58) directly from (2.46), without resorting to the dyad components. Indeed, differentiating (2.46) covariantly and antisymmetrizing, one finds:

$$\varepsilon^{\alpha\beta\gamma\delta}(\phi^B_{;\delta}F^A_{B\beta\gamma} + \phi^A F^B_{A\beta\gamma;\delta}) = \varepsilon^{\alpha\beta\gamma\delta}(\phi^A_{;\beta\gamma\delta} - \phi^A_{;\gamma\beta\delta}) \ . \tag{2.59}$$

Taking into account the antisymmetry property of the tensor $\varepsilon^{\alpha\beta\gamma\delta}$ and the rule for the commutator of the covariant derivatives of mixed quantities having both tensorial and spinorial indices, the right-hand side of the last equation can be cast into the following form:

$$\varepsilon^{\alpha\beta\gamma\delta}(\phi^A_{;\beta\gamma\delta} - \phi^A_{;\gamma\beta\delta}) = \varepsilon^{\alpha\beta\gamma\delta}(\phi^A_{;\beta\gamma\delta} - \phi^A_{;\beta\delta\gamma})$$

$$= \varepsilon^{\alpha\beta\gamma\delta}(\phi^B_{;\beta}F^A_{B\gamma\delta} + \phi^A_{\rho}R^\rho{}_{\beta\gamma\delta}) \ .$$

The second term on the right-hand side vanishes by the cyclic identity of the curvature tensor, and in the first term the antisymmetry property of the tensor density $\varepsilon^{\alpha\beta\gamma\delta}$ can be utilized to yield

$$\varepsilon^{\alpha\beta\gamma\delta}(\phi^A_{;\beta\gamma\delta} - \phi^A_{;\gamma\beta\delta}) = \varepsilon^{\alpha\beta\gamma\delta}(\phi^B_{;\beta}F^A_{B\gamma\delta}) \ . \tag{2.60}$$

Substituting (2.60) in (2.59) completes the proof of (2.58).

2.3.3 *The Riemann tensor*

The discussion of the last subsection has led to the realization that the gauge field $F_a{}^b{}_{\alpha\beta}$ and its associated spinor-tensor $F_A{}^B{}_{\alpha\beta}$ are measures of the curvature. A direct relation between them and the Riemann curvature tensor can be obtained by the substitution of (2.29) in (2.51) and some algebraic manipulations. A simple way to get the same result would be to observe that the covariant parallelism of the connecting quantities $\sigma^\mu{}_{AB'}$, Eq. (2.28), implies

$$\sigma^\mu{}_{AB';\alpha\beta} - \sigma^\mu{}_{AB';\beta\alpha} = 0 \ .$$

Expanding the commutator of the covariant derivatives and making use of the symmetries (2.48), we find

Fundamentals of The SL(2,C) Gauge Theory of Gravitation

$$R^{\alpha}{}_{\beta\gamma\delta}\sigma^{\beta}_{AB'} + F_{A\gamma\delta}^{\ C}\sigma^{\alpha}_{CB'} + F_{B'\alpha\delta}^{\ C'}\sigma^{\alpha}_{AC'} = 0. \tag{2.61}$$

From this equation, with the aid of the algebraic conditions (2.24), it follows readily that

$$F_{A\gamma\delta}^{\ B} = -\frac{1}{2}R^{\alpha}{}_{\beta\gamma\delta}\sigma^{BC'}_{\alpha}\sigma^{\beta}_{AC'} \tag{2.62}$$

and

$$R^{\alpha}{}_{\beta\gamma\delta} = -(F_{A\gamma\delta}^{\ C}\sigma^{\alpha}_{CB'} + F_{B'\gamma\delta}^{\ C'}\sigma^{\alpha}_{AC'})\sigma^{AB'}_{\beta}. \tag{2.63}$$

Passing to the dyad components in the last two equations, the following relations between the gauge field and the Riemann curvature tensor are obtained:

$$F_{a\alpha\delta}^{\ b} = -\frac{1}{2}R^{\alpha}{}_{\beta\gamma\delta}\sigma^{bc'}_{\alpha}\sigma^{\beta}_{ac'} \tag{2.64}$$

and

$$R^{\alpha}{}_{\beta\gamma\delta} = -(F_{a\gamma\delta}^{\ c}\sigma^{\alpha}_{cb'} + F_{b'\gamma\delta}^{\ c'}\sigma^{\alpha}_{ac'})\sigma^{ab'}_{\beta}. \tag{2.65}$$

The Bianchi identities for the spinor-tensor $F_A{}^B{}_{\alpha\beta}$, Eq. (2.58), now follow straightforwardly from (2.64) as a consequence of Bianchi identities for the curvature tensor and Eq. (2.28). As has been shown in the previous subsection, the identities (2.58) lead, in turn, to the Bianchi identities (2.56) for the gauge field. It is, of course, possible to derive the latter directly from (2.64), but the derivation is more complicated since the covariant derivatives of the dyad components $\sigma^{\mu}{}_{ab'}$ of the connecting quantities do not vanish, but rather satisfy the equation

$$\sigma^{\mu}{}_{ab';\nu} = \sigma^{\mu}{}_{cb'}B^{c}_{a\nu} + \sigma^{\mu}{}_{ac'}B^{c'}_{b'\nu}. \tag{2.66}$$

The rest of this subsection will be devoted to finding the decomposition of the gauge field induced by the well-known decomposition of the Riemann curvature tensor into its irreducible components under the MMG, and the ensuing connection with the material content of the physical realm.

Let

$$R_{\alpha\beta} = R^{\rho}{}_{\alpha\rho\beta} \tag{2.67}$$

be the Ricci tensor,

$$R = R^\alpha{}_\alpha \tag{2.68}$$

the Ricci scalar, and $C^\alpha{}_{\beta\gamma\delta}$ the Weyl conformal tensor. Then $R_{\alpha\beta}$ is symmetric in its indices, and $C^\alpha{}_{\beta\gamma\delta}$ is traceless:

$$C^\alpha{}_{\beta\alpha\delta} = 0, \tag{2.69}$$

and is given by

$$C_{\alpha\beta\gamma\delta} = R_{\alpha\beta\gamma\delta} - \frac{1}{2}(g_{\alpha\gamma}R_{\beta\delta} + g_{\beta\delta}R_{\alpha\gamma} - g_{\alpha\delta}R_{\beta\gamma} - g_{\beta\gamma}R_{\alpha\delta})$$

$$+ \frac{1}{6}(g_{\alpha\gamma}g_{\beta\delta} - g_{\alpha\delta}g_{\beta\gamma})R. \tag{2.70}$$

The Weyl tensor, in addition to having a vanishing trace, possesses all the symmetries of the Riemann curvature tensor. It has ten independent components. In view of the Einstein gravitational field equations

$$R_{\alpha\beta} = \kappa(T_{\alpha\beta} - \frac{1}{2}g_{\alpha\beta}T) \tag{2.71}$$

(where $T_{\alpha\beta}$ is the energy-momentum tensor and T is its trace) and (2.70), one has the following representation of the curvature of the space-time manifold:

$$R_{\alpha\beta\gamma\delta} = C_{\alpha\beta\gamma\delta} + \frac{1}{2}(g_{\alpha\gamma}T_{\beta\delta} + g_{\beta\delta}T_{\alpha\gamma} - g_{\alpha\delta}T_{\beta\gamma} - g_{\beta\gamma}T_{\alpha\delta})$$

$$- \frac{1}{3}(g_{\alpha\gamma}g_{\beta\delta} - g_{\alpha\delta}g_{\beta\gamma})T. \tag{2.72}$$

This decomposition leads naturally to the definitions of the following fields in the gauge theory formalism:

$$F^{Wb}_{a\gamma\delta} = -\frac{1}{2}C^\alpha{}_{\beta\gamma\delta}\sigma^{bc'}_\alpha \sigma^\beta_{ac'}, \tag{2.73}$$

and

$$F^{Rb}_{\alpha\gamma\delta} = -\left[\frac{1}{4}\kappa(\delta^{\alpha}{}_{\gamma}T_{\beta\delta} + g_{\beta\delta}T^{\alpha}{}_{\gamma} - \delta^{\alpha}{}_{\delta}T_{\beta\gamma} - g_{\beta\gamma}T^{\alpha}{}_{\delta})\right.$$
$$\left. -\frac{1}{6}\kappa(\delta^{\alpha}{}_{\gamma}g_{\beta\delta} - \delta^{\alpha}{}_{\delta}g_{\beta\gamma})T\right]\sigma^{bc'}_{\alpha}\sigma^{\beta}_{ac'} . \quad (2.74)$$

The first one is the counterpart of the Weyl conformal tensor, while the second is associated through the Ricci tensor, with the energy-momentum. With these definitions the decomposition (2.72) assumes the following form in gauge theory terms:

$$F_{\alpha\beta} = F^W_{\alpha\beta} + F^R_{\alpha\beta} , \quad (2.75)$$

i.e., the gauge field is the sum of the conformal part and the material part.

2.4 Field Equations

2.4.1 Free-field equations

In the previous section a dynamical law, tying up the gauge field with the energy-momentum content of space-time, has been derived via the relation between the gauge field and the Riemann curvature tensor. Thereby the Einstein field equations are reflected in conditions on the gauge field. In order, however, to cast gravity into a bona fide SL(2,C) gauge field theory, the dynamical law ought to be postulated directly in terms of the gauge field itself. We first examine this question in the free-field case.

If the field equations are to be derived from a variational principle, then analogy with the Yang-Mills gauge theory motivates the consideration of the Lagrangian density

$$\mathcal{L}_F = -\frac{1}{4}\sqrt{-g}\, \text{Tr}(F^{\alpha\beta}F_{\alpha\beta}) . \quad (2.76)$$

Here the field $F_{\alpha\beta}$ is given in terms of the potential B_μ by Eq. (2.40). A matrix notation is used here, where the dyad indices, to which the trace operation is applied, are suppressed.

In this notation, the similarity of (2.76) and the electromagnetic Lagrangian density which yields Maxwell's equations is transparent. The variational principle is then given by

$$\delta I = 0 ,$$

with

$$I = \int_\Omega \mathcal{L}_F d^4x ,$$

where $\mathcal{L}_F$ is given by (2.76) and Ω is a four-dimensional volume. The variational variables are the gauge potential matrices B_μ.

The Lagrange equation corresponding to this action is:

$$\partial_\nu(\sqrt{-g}F^{\mu\nu}) - [B_\nu, \sqrt{-g}F^{\mu\nu}] = 0 . \tag{2.77}$$

While these field equations fit naturally into the SL(2,C) gauge theory framework, they do not represent a theory of gravitation consistent with general relativity, as the discussion in the next subsection shows.

2.4.2 Coupling with matter

The way to incorporate matter into the SL(2,C) gauge theory is motivated by the mechanism through which matter and geometry interact within the general relativistic framework. The purely geometrical decomposition (2.70) of the Riemann curvature tensor into the conformal and Ricci parts, has been shown in Subsec. 2.3.3 to induce, via the Einstein field equations, the decomposition

$$F_{\alpha\beta} = F^W_{\alpha\beta} + F^R_{\alpha\beta} \tag{2.78}$$

of the gauge field. The first component F^W is the counterpart of the Weyl conformal tensor, and is given by (2.73), while F^R is a source term, as given by (2.74).

It is evident from (2.78) that Eq. (2.77) cannot be taken as the free-field equations for relativistic gravity, unless one imposes the condition $F^R = 0$. Equation (2.77) then yield the Einstein gravitational free-field equations. Compatibility with the Einstein equations is certainly maintained, in general, if the gauge field is required to satisfy any set of partial differential equations which are equivalent to the Bianchi identities, which, as was shown, when expressed in terms of the gauge field, assume the form (2.56). With the aid of the decomposition (2.78) these equations can be recast in a form more suitable for the viewpoint of gauge theory, viz.,

$$\varepsilon^{\mu\nu\alpha\beta}\{F^W_{\alpha\beta,\nu} - [B_\nu, F^W_{\alpha\beta}]\} = \kappa J^\mu , \qquad (2.79)$$

where

$$J^\mu = -\varepsilon^{\mu\nu\alpha\beta}\{T_{\alpha\beta,\nu} - [B_\nu, T_{\alpha\beta}]\} \qquad (2.80)$$

and

$$T_{\mu\nu a}{}^b = \sigma^\alpha_{ac'}\sigma^{\beta bc'}(g_{[\mu|(\alpha} T_{\beta)|\nu]} + \frac{1}{6}g_{\alpha[\mu}g_{\nu]\beta}T) . \qquad (2.81)$$

Here, J^μ is a 2×2 matrix, interpreted as the current representing the source of the gravitational field.

A Lagrangian density which gives rise to Eq. (2.79) as its Lagrange equation is[13]

$$\mathcal{L}_c = -\frac{1}{2}\varepsilon^{\alpha\beta\mu\nu}F_{\alpha\beta a}{}^b\left\{-\frac{1}{2}F_{\mu\nu b}{}^a + B_{\mu,\nu b}{}^a - B_{\nu,\mu b}{}^a + [B_{\mu b}{}^c, B_{\nu c}{}^a]\right\} . \qquad (2.82)$$

Under the independent variation of the gauge potential B_μ and gauge field $F_{\mu\nu}$ both the relation (2.40) and the Bianchi identities (2.79) are obtained. (For details, see Sec. 3.4.) The quantities $\sigma^\mu{}_{ab'}$ appearing in the field equations are determined by the metric equation (2.28) which is equivalent to

$$\sigma^\mu_{ab';\nu} = B_{\nu a}{}^c \sigma^\mu{}_{cb'} + \sigma^\mu{}_{ac'} B^{\dagger c'}_\nu{}_{b'} , \qquad (2.83)$$

where $B^\dagger$ is the Hermitian conjugate of B.

2.5 Relation to the Newman-Penrose Formalism

The SL(2,C) gauge theory of gravitation is closely related to another mathematical formulation of general relativity due to Newman and Penrose[35]. This section depicts the explicit transformation between the two formalisms, in order to facilitate comparison between results written in the two languages.

All the quantities playing roles in the SL(2,C) gauge theory are given in their dyad components, i.e., their projections along a normalized spinor basis ξ_a^A. Given, then, a spinor basis ξ_a^A subject to the normalization condition

$$\varepsilon_{AB}\xi_a^A\xi_b^B = \varepsilon_{ab} , \qquad (2.84)$$

we form, in terms of the dyad components $\sigma^\mu{}_{ab'}$, a null-tetrad $(l, n, m, \bar{m})$, defined by

$$l^\mu = \sigma^\mu_{00'}, \qquad n^\mu = \sigma^\mu_{11'}, \qquad m^\mu = \sigma^\mu_{01'}. \tag{2.85}$$

The vectors l_μ and n_μ are real, m_μ is complex, and in view of (2.24) the following orthogonality relations hold:

$$l^\mu l_\mu = n^\mu n_\mu = m^\mu m_\mu = \bar{m}^\mu \bar{m}_\mu = 0,$$
$$l^\mu n_\mu = -m^\mu \bar{m}_\mu = 1, \tag{2.86}$$

as well as the completeness relation

$$g^{\mu\nu} = 2[l^{(\mu} n^{\nu)} - m^{(\mu} \bar{m}^{\nu)}]. \tag{2.87}$$

Consider the dyad components of the gauge potential, viz.,

$$B_{ab'} = \sigma^\mu_{ab'} B_\mu. \tag{2.88}$$

To conform with the Newman-Penrose notation for the spin coefficients, denote the elements of these 2×2 matrices by

$$B_{00'} = \begin{bmatrix} \varepsilon & -\kappa \\ \pi & -\varepsilon \end{bmatrix} \qquad B_{01'} = \begin{bmatrix} \beta & -\sigma \\ \mu & -\beta \end{bmatrix}$$
$$B_{10'} = \begin{bmatrix} \alpha & -\rho \\ \lambda & -\alpha \end{bmatrix} \qquad B_{11'} = \begin{bmatrix} \gamma & -\tau \\ \nu & -\gamma \end{bmatrix}. \tag{2.89}$$

Let the dyad components of the Weyl and trace-free Ricci spinors

$$\Psi_{abcd} = \Psi_{ABCD} \xi_a{}^A \xi_b{}^B \xi_c{}^C \xi_d{}^D, \tag{2.90a}$$

$$\phi_{abc'd'} = \phi_{ABC'D'} \xi_a{}^A \xi_b{}^B \bar{\xi}_{c'}{}^{C'} \bar{\xi}_{d'}{}^{D'}, \tag{2.90b}$$

be labelled according to the following scheme:

$$\Psi_0 = -\Psi_{0000}$$

$$\Psi_1 = -\Psi_{0001}$$

$$\Psi_2 = -\Psi_{0011} \tag{2.91}$$

$$\Psi_3 = -\Psi_{0111}$$

$$\Psi_4 = -\Psi_{1111},$$

$$\phi_{00} = -\phi_{000'0'} = \bar\phi_{00}$$

$$\phi_{01} = -\phi_{000'1'} = \bar\phi_{10}$$

$$\phi_{02} = -\phi_{001'1'} = \bar\phi_{20}$$

$$\phi_{11} = -\phi_{010'1'} = \bar\phi_{11} \tag{2.92}$$

$$\phi_{12} = -\phi_{011'1'} = \bar\phi_{21}$$

$$\phi_{22} = -\phi_{111'1'} = \bar\phi_{22}.$$

If finally, the spinor counterpart of the Ricci scalar is denoted by

$$\Lambda = R/24, \tag{2.93}$$

then a straightforward calculation yields the following representation of the gauge field:

$$F_{01'00'} = \begin{bmatrix} \Psi_1 & -\Psi_0 \\ \Psi_2 + 2\Lambda & -\Psi_1 \end{bmatrix}$$

$$F_{11'10'} = \begin{bmatrix} \Psi_3 & -\Psi_2 - 2\Lambda \\ \Psi_4 & -\Psi_3 \end{bmatrix}$$

$$F_{10'00'} = \begin{bmatrix} \phi_{10} & -\phi_{00} \\ \phi_{20} & -\phi_{10} \end{bmatrix}$$

$$\tag{2.94}$$

$$F_{11'01'} = \begin{bmatrix} \phi_{12} & -\phi_{02} \\ \phi_{22} & -\phi_{12} \end{bmatrix}$$

$$F_{11'00'} = \begin{bmatrix} \Psi_2 + \phi_{11} - \Lambda & -\Psi_1 - \phi_{01} \\ \Psi_3 + \phi_{21} & -\Psi_2 - \phi_{11} + \Lambda \end{bmatrix}$$

$$F_{10'01'} = \begin{bmatrix} -\Psi_2 + \phi_{11} + \Lambda & \Psi_1 - \phi_{01} \\ -\Psi_3 + \phi_{21} & \Psi_2 - \phi_{11} - \Lambda \end{bmatrix}.$$

Now consider Eq. (2.40b), relating the gauge field with the gauge potential. Contracting with σ^μ yields

$$\partial_{cd'} B_{ab'} - \partial_{ab'} B_{cd'} - (\nabla_{cd'} \sigma^\mu_{ab'} - \nabla_{ab'} \sigma^\mu_{cd'}) B_\mu$$

$$+ [B_{ab'}, B_{ad'}] = F_{ab'cd'} \tag{2.95}$$

where the dyad components

$$\partial_{ab'} = \sigma^\mu_{ab'} \partial_\mu$$

$$\nabla_{ab'} = \sigma^\mu_{ab'} \nabla_\mu \tag{2.96}$$

of the partial and covariant derivatives are used. The term with the derivatives of the σ's can be eliminated with the aid of (2.66), leading to the equation

$$\partial_{cd'} B_{ab'} - \partial_{ab'} B_{cd'} - (B_{cd'})_a^{\ f} B_{fb'} - (B^\dagger_{d'c})^{f'}_{\ b'} B_{af'}$$

$$+ (B_{ab'})_c^{\ f} B_{fd'} + (B^\dagger_{b'a})^{f'}_{\ d'} B_{cf'} + [B_{ab'}, B_{cd'}] = F_{ab'cd'}. \tag{2.97}$$

Likewise, Eq. (2.56) is equivalent to

$$\partial_{ab'} F_{cd'ef'} + \partial_{cd'} F_{ef'ab'} + \partial_{ef'} F_{ab'cd'} - (B_{ab'})_c^{\ g} F_{gd'ef'}$$

$$- (B^\dagger_{b'a})^{g'}_{\ d'} F_{cg'ef'} - (B_{ab'})_c^{\ g} F_{cd'gf'} - (B^\dagger_{b'a})^{g'}_{\ f'} F_{cd'eg'}$$

$$- (B_{cd'})_e^{\ g} F_{gf'ab'} - (B^\dagger_{d'c})^{g'}_{\ f'} F_{cg'ab'} - (B_{cd'})_a^{\ g} F_{ef'gb'}$$

$$- (B^\dagger_{d'c})^{g'}_{\ b'} F_{ef'ag'} - (B_{ef'})_a^{\ g} F_{gb'cd'} - (B^\dagger_{f'e})^{g'}_{\ b'} F_{ag'cd'}$$

$$- (B_{ef'})_c^{\ g} F_{ab'gd'} - (B^\dagger_{f'e})^{g'}_{\ d'} F_{ab'cg'}$$

$$= [B_{ab'}, F_{cd'ef'}] + [B_{cd'}, F_{ef'ab'}] + [B_{ef'}, F_{ab'cd'}]. \qquad (2.98)$$

Finally, a straightforward calculation yields for the metric equation (2.83) the following:

$$\partial_{ab'}\sigma^{\mu}_{cd'} - \partial_{cd'}\sigma^{\mu}_{ab'} = (B_{ab'})_c{}^f \sigma^{\mu}_{fd'} - (B_{cd'})_a{}^f \sigma^{\mu}_{fb'}$$

$$+ \sigma^{\mu}_{cf'}(B^{\dagger}_{b'a})^{f'}{}_{d'} - \sigma^{\mu}_{af'}(B^{\dagger}_{d'c})^{f'}{}_{b'}.$$

Equations (2.97) and (2.98) along with the metric equation are precisely the combined set of the Newman-Penrose equations.

2.6 Derivation by Palatini's Method

In Subsec. 2.4.2 the dynamical laws of the SL(2,C) gauge theory of gravitation have been expressed as a set of partial differential equations combined with algebraic constraints relating components of the gauge field to the energy-momentum content of space-time. The differential equations are the Lagrange equations derived from a variational principle based on the Lagrangian density (2.81), whereas the algebraic conditions are supplementary requirements, independent of the variational principle. Furthermore, the definition (2.32) of the gauge potential seems to be an act of serendipity, bearing no intrinsic relation to the other postulates underlying the formalism.

In this section a unified derivation will be presented, whereby both the Einstein field equations of gravity and the connection between the gauge potential and the null-tetrad emerge from a single variational principle (to augment the variational principle of Sec. 2.4). The method used here is an extension to that of Palatini[20].

It will be useful to introduce the notation

$$S_a{}^b{}_{\mu\nu} = \frac{1}{2}(\sigma_{\mu ac'}\sigma_\nu{}^{bc'} - \sigma_{\nu ac'}\sigma_\mu{}^{bc'}). \qquad (2.99)$$

By employing, repeatedly, the anti-commutation relations of the σ's (see Subsec. 2.2.3 following Eq.(2.23)), it can be verified that the following "completeness relations" hold:

$$\text{Tr}(S^{\mu\nu}S_{\alpha\beta}) = \frac{1}{2}(\delta^\mu_\beta \delta^\nu_\alpha - \delta^\mu_\alpha \delta^\nu_\beta) + \frac{1}{2}i\epsilon^{\mu\nu}{}_{\alpha\beta}, \qquad (2.100)$$

where ϵ is the Levi-Civita antisymmetric tensor.

Consider now the simplest Lagrangian density that can be constructed of the elements of the SL(2,C) gauge theory, which is linear in the gauge field, viz.,

$$\mathcal{L}_0 = -2\sigma \, \text{Tr}(S^{\mu\nu}F_{\mu\nu}) \,, \tag{2.101}$$

where

$$\sigma = \sqrt{-g} \,. \tag{2.102}$$

This Lagrangian density is interpreted as a concomitant of the null-tetrad $\sigma^\mu{}_{ab'}$, the gauge potential B_μ, and the first order derivatives of the gauge potential. Thus σ in Eq. (2.101) is written in terms of the metric tensor $g_{\mu\nu}$, but the latter is taken as the tensor concomitant of the null-tetrad defined by

$$g_{\mu\nu} = \sigma_{\mu ab'}\sigma_\nu{}^{ab'} \,. \tag{2.103}$$

We have, therefore,

$$\mathcal{L}_0 = \mathcal{L}_0(\sigma^\mu_{ab'}, (B_\mu)_a{}^b, (B_{\mu,\nu})_a{}^b) \,. \tag{2.104}$$

To evaluate $\mathcal{L}_0$ at a given point in the space-time manifold, substitute in (2.101) the expression (2.64) of the gauge field in terms of the Riemann tensor, which can be cast in the form

$$F_{\mu\nu} = \frac{1}{2}S^{\alpha\beta}R_{\mu\nu\alpha\beta} \,, \tag{2.105}$$

(with dyad indices suppressed). In view of the completeness relations (2.100) and the cyclic identity satisfied by the curvature tensor, one easily finds:

$$\mathcal{L}_0 = \sqrt{-g}R \,. \tag{2.106}$$

While $\mathcal{L}_0$ and the Ricci scalar density $\sqrt{-g}R$ have the same numerical value, their dependence on their respective arguments are obviously different. In particular, it is worth noticing that the form (2.101) embodies an inherent coupling between the space-time and the internal spinor-space structures[43].

In parallel to the Palatini method[44] as applied to the Lagrangian density $\sqrt{-g}R$, whereby the metric tensor and the affine connections are treated as (*a priori*) independent variables, we now turn to the Lagrangian $\mathcal{L}_0$ defined by

Fundamentals of The SL(2,C) Gauge Theory of Gravitation

(2.101). In view of (2.104) we consider the null tetrad $\sigma^\nu{}_{ab'}$ and the gauge potential B_μ as independent variables, i.e., the relation (2.32) is not assumed. The variation in $\mathcal{L}_0$ induced by variation of $\sigma^\mu{}_{ab'}$ is easily found to be

$$\delta\mathcal{L}_0 = (\sqrt{-g} F^a_{c\mu\nu} \sigma^{\nu cb'} - \frac{1}{2}\mathcal{L}_0 \sigma^{ab'}_\mu)\delta\sigma^\mu_{ab'} , \qquad (2.107)$$

so that the associated Lagrange equation, in view of (2.106), is given by

$$F^a_{c\mu\nu}\sigma^{\nu cb'} - \frac{1}{2}R\sigma^{ab'}_\mu = 0 . \qquad (2.108)$$

The physical content of this equation is found upon multiplying it by $\sigma^\nu{}_{ab'}$. Simple manipulations, taking into account (2.64) and the completeness relations (2.100), yield:

$$R_{\mu\nu} - \frac{1}{2}g_{\mu\nu}R = 0 , \qquad (2.109)$$

i.e., the vacuum Einstein field equations.

A more tedious, but straightforward, calculation yields the Lagrange equation associated with the variation of B_μ, viz.,

$$A^{\mu\nu}{}_{;\nu} + [A^{\mu\nu}, B_\nu] = 0 , \qquad (2.110)$$

which turns out to be equivalent to Eq. (2.32) defining the gauge potential in terms of the null tetrad.

Finally, to obtain the Einstein equations in the presence of matter, the Lagrangian density (2.101) has to be replaced by

$$\mathcal{L} = \mathcal{L}_0 - 2\kappa\mathcal{L}_M , \qquad (2.111)$$

where $\mathcal{L}_M$ is the matter Lagrangian density, depending on the gauge potential (along with the physical fields) but not on its derivatives. (For more details the reader is referred to Ref. 20.)

3
QUADRATIC LAGRANGIANS

El ser de todos los seres
solo formó la unidá;
lo demás lo ha criado el hombre
después que aprendió a contar

Jose Hernandez, *Martin Fierro*,
part II, line 4308 (1872).

3.1 *Linear and Quadratic Lagrangians*

In this chapter we review recent developments in the theory of quadratic-Lagrangian formulation of general relativity theory. Of particular interest will be the quadratic invariants obtained from the Riemann tensor, and the question whether they are suitable to serve as Lagrangians.

The discussion starts, in the next section, with Hilbert's familiar Lagrangian. The latter is linear in the Riemann tensor and historically served as the basis for the variational formulation of general relativity theory. Hilbert's original approach took the components of the metric tensor and their derivatives as the independent field variables. Subsequently, Palatini used the same Lagrangian but performed the variations on a larger number of independent variables to obtain, in addition to Einstein's field equations, the affine connections in terms of the metric tensor and its derivatives. The extension of Hilbert's Lagrangian due to Palatini will serve as a model for our approach. In an analogous manner we will extend the Lanczos invariant, which is spurious for variational purpose when the metric tensor components are taken as the independent field variables, into a Lagrangian which yields Einstein's field equations.

In Sec. 3.3 the quadratic invariants which can be constructed from the Riemann tensor are reviewed, including the Lanczos invariant. In Sec. 3.4 we discuss Carmeli's quadratic Lagrangian for the SL(2,C) gauge theory of gravitation. This is followed by the exhibition of a one-parameter family of quadratic Lagrangians for the SL(2,C) gauge theory of gravitation in Sec. 3.5. Only one Lagrangian of this family turns out to be SL(2,C) invariant, and it is, in fact, identical in its extremal value to Carmeli's Lagrangian[26]. In Sec. 3.6 we discuss a quadratic Lagrangian due to Nissani, which produces Einstein's gravitational field equations in their tensorial form[28,29]. The last section is devoted to a summary of the chapter.

3.2 Hilbert's Lagrangian

As has been pointed out before, Hilbert[32] obtained the gravitational field equations of general relativity theory from a variational principle. To this end the action integral is given by

$$I = \int_\Omega (\mathcal{L}_G - 2\kappa\mathcal{L}_M)d^4x , \qquad (3.1)$$

with $\mathcal{L}_G = \sqrt{-g}R$, where R is the Ricci scalar, $\mathcal{L}_M$ is the matter Lagrangian density and Ω is a four-dimensional domain.

The scalar density $\mathcal{L}_G$, describing the gravitational part of Hilbert's Lagrangian, is given by

$$\mathcal{L}_G = \sqrt{-g}R = \sqrt{-g}g^{\mu\nu}(\Gamma^\rho_{\mu\nu,\rho} - \Gamma^\rho_{\mu\rho,\nu} + \Gamma^\sigma_{\mu\nu}\Gamma^\rho_{\rho\sigma} - \Gamma^\sigma_{\mu\rho}\Gamma^\rho_{\nu\sigma}) , \qquad (3.2)$$

where $\Gamma^\mu{}_{\alpha\beta}$ are the Christoffel symbols,

$$\Gamma^\gamma_{\alpha\beta} = \frac{1}{2}g^{\gamma\rho}(g_{\alpha\rho,\beta} + g_{\beta\rho,\alpha} - g_{\alpha\beta,\rho}) , \qquad (3.3)$$

i.e., functions of the metric tensor and its derivatives. A straightforward calculation gives

$$\delta\int \mathcal{L}_G d^4x = \int \sqrt{-g}G^{\mu\nu}\delta g_{\mu\rho}d^4x , \qquad (3.4a)$$

where $G^{\alpha\beta}$ is the Einstein tensor

Quadratic Lagrangians

$$G^{\alpha\beta} = R^{\alpha\beta} - \frac{1}{2}g^{\alpha\beta}R,$$

and

$$\delta \int \mathcal{L}_M d^4x = \frac{1}{2} \int \sqrt{-g}\, T^{\alpha\beta} \delta g_{\alpha\beta} d^4x, \qquad (3.4b)$$

where $T^{\alpha\beta}$ is the energy-momentum tensor of the material physical systems described by the Lagrangian density $\mathcal{L}_M$. From this action principle one obtains Einstein's gravitational field equations

$$G^{\mu\nu} = \kappa T^{\mu\nu}. \qquad (3.5)$$

Palatini[44] suggested a "stronger" variation on the same Lagrangian of Hilbert, namely, a variation on a larger number of independent variables. In Palatini's approach one assumes that the metric components and the functions Γ (affine connections) are independent. Consequently, Palatini's Lagrangian is a function of 50 variables (10 components of the metric tensor and 40 of the Γ's, the latter are assumed to be symmetric). Hilbert's Lagrangian, therefore, can be considered as a projection of Palatini's Lagrangian on the 10-dimensional "hypersurface" defined by Eq. (3.3). This "hypersurface" defines the domain of existence of Hilbert's Lagrangian, and the extremal values of Palatini's action are located on it.

In Hilbert's approach the affine connections are determined *a priori*. On the other hand, in Palatini's approach the relation between the connections and the metric is deduced from the action principle; Eq. (3.3) now expresses a dynamical law (rather than a mathematical definition). In Sec. 3.6 we will apply a similar procedure on Lanczos' invariant, ignoring the functional relation between Weyl's conformal tensor and the metric tensor. In this way we obtain a Lagrangian with a larger number of independent variables, which yields Einstein's field equations.

In the next section we review the quadratic invariants which can be constructed from the Riemann tensor, one of them being related to the Lagrangian of Sec. 3.6.

3.3 The Quadratic Invariants

As has been mentioned before, our experience from various fields of physics, usually having a small number of invariants, shows that there are no invariants devoid of physical significance. Contrary to this, the general theory of relativity provides a large number of invariants constructed out of its field functions, most of them seem to have no obvious physical meaning. However, some of these invariants are used as variational bases for theories of gravitation other than Einstein's general relativity. In this section we focus our attention to several of these invariants, including one which is closely related to the quadratic Lagrangian to be discussed in Sec. 3.6.

Of particular interest are the following scalar densities[45–48]:

$$\mathcal{L}_{L1} = \sqrt{-g}R^2 \,,$$

$$\mathcal{L}_{L2} = \sqrt{-g}R_{\alpha\beta}R^{\alpha\beta} \,,$$

$$\mathcal{L}_{L3} = \sqrt{-g}R^{\alpha\beta\gamma\delta}R_{\alpha\beta\gamma\delta} \,, \qquad (3.6)$$

$$\mathcal{L}_{L4} = \varepsilon^{\mu\nu\rho\sigma}R^{\alpha\beta}{}_{\rho\sigma}R_{\alpha\beta\mu\nu} \,,$$

$$\mathcal{L}_{L5} = \varepsilon^{\mu\nu\rho\sigma}R_{\alpha\beta\rho\sigma}R_{\gamma\delta\mu\nu}\varepsilon^{\alpha\beta\gamma\delta} \,.$$

The first three densities in Eqs. (3.6) yield, via variational calculus, the field equations of theories which are alternatives to general relativity and are known as the Weyl-Eddington-type theories of gravitation[49,50]. The field equations of these theories are differential equations of the fourth order, in contrast to Einstein's second-order equations. As such they are also known as "higher-order theories". This fact is considered as one of the deficiencies of these theories as compared with Einstein's theory, since among other things they do not reduce to Poisson's equation in the Newtonian limit.

Nevertheless, the higher-order theories, like Einstein's theory, do admit the spherically symmetric Schwarzschild solution in vacuum. In these theories of gravitation, however, the Birkhoff theorem on the uniqueness of the Schwarzschild solution is not valid[45]. Unfortunately, all the experimental verifications of Einstein's theory of gravitation are based on the Schwarzschild solution. This fact, as has been pointed out by Havas, renders it impossible to distinguish experimentally between the various theories[45].

As to the last two scalar densities in Eqs. (3.6), Lanczos[46] proved that they are spurious as Lagrangians when considered as functions of the metric and its derivatives. An application of the variational calculus, with respect to the metric components, to these two scalar densities $\mathcal{L}_{L4}$ and $\mathcal{L}_{L5}$, yields Lagrange

Quadratic Lagrangians

expressions which vanish identically. In Sec. 3.6 it will be shown that $\mathscr{L}_{L4}$ can be extended into a Lagrangian which produces the Einstein field equations.

In the next section we discuss Carmeli's quadratic Lagrangian which generates the Newman-Penrose equations, the latter being equivalent to the Einstein equations. For the sake of completeness, we will also recapitulate some aspects of the topics treated in Chapter 2 so far as they are relevant to the Lagrangian approach.

3.4 Carmeli's Lagrangian

The SL(2,C) gauge theory of gravitation due to Carmeli, referred to in the last chapter, is equivalent to Einstein's theory of gravitation, and is derived from a Lagrangian which is quadratic in the gauge field[13],

$$\mathscr{L}_C = -\frac{1}{2} \varepsilon^{\alpha\beta\mu\nu} F_{\alpha\beta a}{}^b \left[-\frac{1}{2} F_{\mu\nu b}{}^a + B_{\mu,\nu b}{}^a - B_{\nu,\mu b}{}^a + [B_{\mu b}{}^c, B_{\nu c}{}^a] \right] \qquad (3.7)$$

(The gauge field $F_{\alpha\beta a}{}^b$ and gauge potential $B_{\alpha a}{}^b$ were defined in Sec. 2.3.) This Lagrangian is a scalar density with respect to the general coordinate transformations, and is invariant under the transformations of the gauge group of the theory.

For the purpose of variational calculus, one considers the matrix elements of the gauge field tensor and the potential vector as the independent field variables. In this way one obtains from the variation of the gauge field matrix elements,

$$\frac{\partial \mathscr{L}_C}{\partial F_{\mu\nu a}{}^b} = 0 , \qquad (3.8)$$

namely,

$$F_{\mu\nu} = B_{\mu,\nu} - B_{\nu,\mu} + [B_\mu, B_\nu] . \qquad (3.9)$$

This equation is precisely the definition (2.40) of the gauge field, which, in the present version, is a dynamical law rather than given *a priori*.

From the variation of the matrix elements of the gauge potential vector B_μ, one obtains the Bianchi identities in their diadic form [Eq. (2.56)], which are part of the Newman-Penrose field equations,

$$\varepsilon^{\mu\nu\alpha\beta}\left[F_{\alpha\beta,\nu} - [B_\nu, F_{\alpha\beta}]\right] = 0 . \tag{3.10}$$

Substituting Eq. (3.9) in Eq. (3.7) then gives

$$\mathscr{L}_C^E = -\frac{1}{4}\operatorname{Tr}(\varepsilon^{\mu\nu\alpha\beta}F_{\alpha\beta}F_{\mu\nu}) , \tag{3.11}$$

in which (and in the sequel) the superscript E indicates the extremal value.

As was shown in Sec. 2.3, the decomposition of the Riemann curvature tensor into its irreducible components induces a decomposition of the gauge field,

$$F_{\mu\nu} = F_{\mu\nu}^W + F_{\mu\nu}^R , \tag{3.12}$$

where W and R denote the Weyl and the Ricci parts. If this is substituted in the Lagrangian (3.7), with F^R replaced by its energy-momentum tensorial expression (via the Einstein field equations), then the Lagrange equations (3.10) and (3.9), associated with the Lagrangian (3.7), are the Newman-Penrose field equations in the presence of matter, which are equivalent to Einstein's field equations[35].

The Lagrangian form (3.11), along with Eq. (2.64), point to the close relationship between Carmeli's Lagrangian and Lanczos' invariants. In the next section we prove, within the framework of the SL(2,C) gauge theory of gravitation, the existence of a one-parameter family of Lagrangians, with only one of which being invariant under the group SL(2,C). This invariant Lagrangian is equal, in its extremal value, to Carmeli's Lagrangian.

3.5 The Lagrangian Family

In the last section we introduced Carmeli's Lagrangian, which is quadratic in the SL(2,C) gauge field of gravitation and hence is quadratic in the Riemann curvature tensor. In this section we show, following Nissani[26], the existence of a one-parameter family of Lagrangians.

Consider the following one-parameter expression:

$$\mathscr{L}_\kappa = \operatorname{Tr}[-(*L^{\alpha\beta}L_{\alpha\beta} + *K^{\alpha\beta}K_{\alpha\beta}) + 4(*L^{\alpha\beta}B_{\alpha,\beta} + B_\alpha B_\beta *K^{\alpha\beta})$$
$$+ k\varepsilon^{\alpha\beta\gamma\delta}B_\alpha B_\beta B_{\gamma,\delta}] , \tag{3.13}$$

Quadratic Lagrangians 35

where L and K are undefined antisymmetric tensors whose components are 2×2 matrices, B is the gauge potential vector, k is an arbitrary complex parameter, and * denotes the duality operator. By definition, $\mathscr{L}_\kappa$ is a scalar density with respect to the general coordinate transformations in space-time, but is not necessarily invariant under the SL(2,C) group transformations.

From the variation of the matrix elements of $L_{\alpha\beta}$ we derive the relationship between the tensor density $*L$ and the gauge potential vector B:

$$*L^{E\,\gamma\delta} = \varepsilon^{\gamma\delta\alpha\beta} B_{\alpha,\beta} \; . \tag{3.14}$$

From this equation one gathers that L, in its extremal value, is identical with the rotor part of the gauge field. Similarly, varying the matrix elements of $K_{\alpha\beta}$ yields the following relationship between the tensor density $*K$ and the gauge potential vector B:

$$*K^{E\,\gamma\delta} = \varepsilon^{\gamma\delta\alpha\beta} B_\alpha B_\beta \; . \tag{3.15}$$

Thus, the tensor K, in its extremum, is identical with the commutator part of the gauge field. Finally, from the variation of the matrix elements of B_α we obtain:

$$*L^{E\,\alpha\beta}{}_{,\beta} = [B_\beta, *K^{E\,\alpha\beta}] \; . \tag{3.16}$$

One can easily verify that Eqs. (3.14)–(3.16) are equivalent to Eqs. (3.9) and (3.10) derived from Carmeli's Lagrangian.

Substituting in the Lagrangian family (3.13) the extremal values of the tensor L and K, and assigning the value 4 to the parameter k, we arrive at an SL(2,C) invariant expression, which is identical with Carmeli's Lagrangian.

3.6 The Tensorial Quadratic Lagrangian

In the previous sections Carmeli's quadratic Lagrangian, which gives the Newman-Penrose version of the Einstein gravitational field equations, was introduced. We also proved the existence, in the spinor space, of a Lagrangian family, with one of its members being an SL(2,C) invariant. This member is equal, in its extremal value, to Carmeli's Lagrangian.

In this section we discuss a purely tensorial quadratic Lagrangian due to Nissani[28,29], which gives the Einstein gravitational field equations in their classical (tensorial) form without resorting to the spinor space. This Lagrangian is identical, in its extremal value, with the Lanczos scalar density $\mathscr{L}_{L4}$ given in Eqs. (3.6).

3.6.1 *General properties*

This subsection is devoted to preliminaries needed for the introduction of the tensorial quadratic Lagrangian. The proofs of the statements made here will be given in the following subsections.

We begin by defining three 4-index tensors. The first is the Riemann tensor, which is a function of the metric tensor and its derivatives,

$$R^\alpha_{\beta\gamma\delta}(g_{\alpha\beta}, g_{\alpha\beta,\gamma}, g_{\alpha\beta,\gamma\delta}) = \Gamma^\alpha_{\beta\delta,\gamma} - \Gamma^\alpha_{\beta\gamma,\delta} + \Gamma^\alpha_{\mu\gamma}\Gamma^\mu_{\beta\delta} - \Gamma^\alpha_{\mu\delta}\Gamma^\mu_{\beta\gamma}, \qquad (3.17)$$

where Γ are the Christoffel symbols,

$$\Gamma^\alpha_{\beta\gamma} = \frac{1}{2}g^{\alpha\mu}(g_{\beta\mu,\gamma} + g_{\gamma\mu,\beta} - g_{\beta\gamma,\mu}). \qquad (3.18)$$

The following identities

$$\varepsilon^{\alpha\beta\gamma\delta}R^\mu_{\beta\gamma\delta} = 0, \qquad (3.19)$$

$$\varepsilon^{\alpha\beta\gamma\delta}R_{\mu\nu\gamma\delta;\beta} = 0, \qquad (3.20)$$

satisfied by the Riemann tensor, will be used in the sequel.

The second tensor is a function of the metric tensor and the energy-momentum tensor $T^{\alpha\beta}$, and is defined by

$$U_{\alpha\beta\gamma\delta}(g_{\alpha\beta}, T_{\alpha\beta}) = T_{\beta[\delta}g_{\gamma]\alpha} + T_{\alpha[\gamma}g_{\delta]\beta} + \frac{2}{3}g_{\alpha[\delta}g_{\gamma]\beta}T, \qquad (3.21)$$

where square brackets denote antisymmetrization, and T is the trace of the energy-momentum tensor. It is easily seen that Eq. (3.21) yields

$$U^\alpha_{\beta\alpha\delta} = T_{\beta\delta} - \frac{1}{2}g_{\beta\delta}T. \qquad (3.22)$$

The Third 4-index tensor, denoted by $C_{\alpha\beta\gamma\delta}$ and called the gravitational tensor, will be left unspecified, with no condition imposed on it *a priori*. Its meaning and content will follow from the requirement of the extremum of action.

With the aid of these three tensors we now define four scalar densities,

Quadratic Lagrangians

$$\mathcal{L}_1(C_{\alpha\beta\gamma\delta}; g_{\alpha\beta}) = \sqrt{-g}\, C_{\alpha\beta[\gamma\delta]} C_{\mu\nu[\rho\sigma]} g^{\alpha\gamma} g^{\mu\rho} g^{\beta\nu} g^{\delta\sigma}\,, \tag{3.23}$$

$$\mathcal{L}_2(C_{\alpha\beta\gamma\delta}; g_{\alpha\beta}) = \frac{1}{4} \varepsilon^{\gamma\delta\rho\sigma} C_{\alpha\beta\gamma\delta} C_{\mu\nu\rho\sigma} g^{\alpha\nu} g^{\beta\mu}\,, \tag{3.24}$$

$$\mathcal{L}_3(C_{\alpha\beta\gamma\delta}; g_{\alpha\beta}, g_{\alpha\beta,\gamma}, g_{\alpha\beta,\gamma\delta}) = -\frac{1}{2} \varepsilon^{\gamma\delta\rho\sigma} C_{\alpha\beta\gamma\delta} R^{\beta}{}_{\nu\rho\sigma} g^{\alpha\nu}\,, \tag{3.25}$$

$$\mathcal{L}_4(C_{\alpha\beta\gamma\delta}; T_{\alpha\beta}; g_{\alpha\beta}) = \frac{\kappa}{2} \varepsilon^{\gamma\delta\rho\sigma} U_{\alpha\beta\gamma\delta} C_{\mu\nu\rho\sigma} g^{\alpha\nu} g_{\beta\mu}$$

$$= \kappa \varepsilon^{\rho\sigma\delta[\mu} g^{\nu]\beta} C_{\mu\nu\rho\sigma} T_{\beta\delta} + \frac{\kappa}{3} \varepsilon^{\mu\nu\rho\sigma} C_{\mu\nu\rho\sigma} T\,. \tag{3.26}$$

(κ is Einstein's gravitational constant.)

Notice that the first three scalar densities are functions of the gravitational variables alone, while the fourth scalar density describes the interaction between the gravitational field and matter. It is, therefore, natural to assume that the sum of the first three scalar densities,

$$\mathcal{L}_F = \sum_{1}^{3} \mathcal{L}_i\,, \tag{3.27}$$

constitutes the vacuum space-time Lagrangian. Indeed, it leads, as will be shown in the sequel, to Einstein's field equations in vacuum,

$$R_{\alpha\beta} = 0\,, \tag{3.28}$$

where

$$R_{\alpha\beta} = R^{\mu}{}_{\alpha\mu\beta}\,, \tag{3.29}$$

is the Ricci tensor.

Adding the interaction (between the gravitational field and matter) term $\mathcal{L}_4$ gives the total Lagrangian:

$$\mathcal{L} = \sum_{1}^{4} \mathcal{L}_i\,, \tag{3.30}$$

from which one obtains, as we will see, the Einstein gravitational field equations in the presence of matter,

$$R_{\alpha\beta} = \kappa(T_{\alpha\beta} - \frac{1}{2}g_{\alpha\beta}T) \,. \tag{3.31}$$

It is worthwhile mentioning that one can get rid of the second derivatives of the metric tensor, appearing in the scalar density $\mathcal{L}_3$, by adding a divergence term,

$$\mathcal{L}'_3 = \varepsilon^{\mu\nu\rho\sigma}(C^{\alpha}{}_{\beta\rho\sigma}\Gamma^{\beta}_{\alpha\nu})_{,\mu} \,. \tag{3.32}$$

It follows that the Lagrange equation obtained from the total Lagrangian will include no terms with derivatives of higher order than the second. It should be emphasized, however, that since $\mathcal{L}'_3$ is not tensorial, the Lagrangian in its first-order version is no longer a scalar density.

It will be noted that although no assumptions have been made about the symmetry of the gravitational tensor C, its symmetric part with respect to its last two indices does not contribute to the Lagrangian. Hence, with no loss of generality, we assume that the tensor C, in its extremal value, is skew-symmetric in the last pair of indices,

$$C^E_{\alpha\beta\gamma\delta} = -C^E_{\alpha\beta\delta\gamma} \,. \tag{3.33}$$

There exist three more scalar densities which can be added to the total Lagrangian (3.30) without changing the resulting Lagrange equations, and they are of interest for our analysis. The first one describes the matter, and is given by

$$\mathcal{L}_5 = \frac{\kappa^2}{4}\varepsilon^{\gamma\delta\rho\sigma}U_{\alpha\beta\gamma\delta}U_{\mu\nu\rho\sigma}g^{\alpha\nu}g^{\beta\mu} = \frac{\kappa^2}{2}\varepsilon^{\gamma\delta\rho\sigma}T_{\rho\delta}T_{\gamma\sigma} \,, \tag{3.34}$$

and its absence from the total Lagrangian is to be noticed. This term and its variational derivative $\delta\mathcal{L}_5/\delta T_{\alpha\beta}$ vanish in the extremal value because of the symmetry of the energy-momentum tensor,

$$T_{\alpha\beta} = T_{\beta\alpha} \,, \tag{3.35}$$

which is a consequence of the Einstein field equations. Hence the addition of such a term does not contribute to the extremal value of the action.

Quadratic Lagrangians

The second density is given by

$$\mathcal{L}_6 = -\frac{\kappa}{2} \varepsilon^{\gamma\delta\rho\sigma} U_{\alpha\beta\gamma\delta} R^{\beta}{}_{\nu\rho\sigma} g^{\alpha\nu}$$

$$= -\kappa \varepsilon^{\rho\sigma\delta[\mu} g^{\nu]\gamma} R_{\mu\nu\rho\sigma} T_{\gamma\delta} - \frac{\kappa}{3} \varepsilon^{\mu\nu\rho\sigma} R_{\mu\nu\rho\sigma} T . \quad (3.36)$$

It vanishes identically as a consequence of Eq. (3.19), and therefore this term too does not contribute to the Lagrangian. The relevance of $\mathcal{L}_6$ is due to the fact that it does not vanish identically if one assumes Γ to be independent of g. Its addition to the total Lagrangian permits us to consider the variations of g and Γ as independent.

The third scalar density is proportional to the Lanczos invariant $\mathcal{L}_4$ given by Eqs. (3.6):

$$\mathcal{L}_7 = A \varepsilon^{\gamma\delta\rho\sigma} R^{\alpha}{}_{\beta\gamma\delta} R^{\beta}{}_{\alpha\rho\sigma} , \quad (3.37)$$

where A is an arbitrary constant. It is considered as a function of the metric and its derivatives, and hence, according to Lanczos, its Euler-Lagrange expressions vanish identically. Adding this term, therefore, cannot affect the Lagrange equations obtained from the total Lagrangian (3.30).

In the extremal value of the action, the total Lagrangian (3.30) can be shown to be proportional to the Lanczos invariant, and assumes the form

$$\mathcal{L}^E = -\frac{1}{4} \varepsilon^{\alpha\beta\gamma\delta} R^{\nu}{}_{\mu\alpha\beta} R^{\mu}{}_{\nu\gamma\delta} . \quad (3.38)$$

Consequently, the addition of $\mathcal{L}_7$ with $A = \frac{1}{4}$ yields a vanishing action in its extremal value.

3.6.2 *The Variation of $\mathcal{L}$*

From the variation in the total Lagrangian (3.30) of the components of the gravitational tensor C, the energy-momentum tensor T, and the metric tensor g, one obtains a system of 122 Lagrange equations for 122 (96 + 16 + 10) functions. For a given metric tensor, the Lagrange equations become a system of algebraic equations, of which 112 are linear and 10 are quadratic, relating the 112 functions, C and T, to the metric tensor components. In this case, therefore, the number of equations exceeds the number of unknown functions by 10.

The resulting system of Lagrange equations reduces to the following set of equations:

$$C^E_{\alpha\beta\gamma\delta} + \kappa U^E_{\alpha\beta\gamma\delta} = R^E_{\alpha\beta\gamma\delta}, \tag{3.39}$$

$$C^{E\gamma}{}_{\beta\gamma\delta} = 0. \tag{3.40}$$

By contracting the indices α and γ in Eq. (3.39), and using Eqs. (3.40) and (3.22), we obtain the Einstein gravitational field equations in the presence of matter, Eq. (3.31). This equation, in turn, entails the symmetry of the energy-momentum tensor at its extremal value. Moreover, the symmetry of the energy-momentum tensor and the definition (3.21) of the 4-index energy-momentum tensor imply:

$$\varepsilon^{\alpha\nu\rho\sigma} U^E_{\beta\nu\rho\sigma} = 0, \tag{3.41}$$

which, along with Eqs. (3.19) and (3.39), gives

$$\varepsilon^{\alpha\nu\rho\sigma} C^E_{\beta\nu\rho\sigma} = 0. \tag{3.42}$$

Equations (3.39) and (3.40), in view of the definition (3.21), then lead to the identification of the gravitational tensor $C_{\alpha\beta\gamma\delta}$, in its extremal value, with Weyl's conformal tensor, which is the motivation for the notation and title assigned to it. In the following we prove that any solution of Einstein's field equations is a solution of the Lagrange equations obtained by varying the components of the tensors $C_{\alpha\beta\gamma\delta}$, $T_{\alpha\delta}$, and $g_{\alpha\beta}$.

3.6.3 *The Variation of $C_{\alpha\beta\gamma\delta}$ and $T_{\alpha\beta}$*

From the variation of the components of the gravitational tensor in the total Lagrangian (3.30) we obtain the following 96 Lagrange equations:

$$\sqrt{-g}\, C^E_{\mu\nu\rho\sigma} g^{\alpha[\gamma} g^{\delta]\sigma} g^{\beta\nu} g^{\mu\rho}$$
$$+ \frac{1}{2} \varepsilon^{\gamma\delta\rho\sigma} (C^{E\beta}{}_{\nu\rho\sigma} + \kappa U^{E\beta}{}_{\nu\rho\sigma} - R^{E\beta}{}_{\nu\rho\sigma}) g^{\alpha\nu} = 0. \tag{3.43}$$

It is easily verified that any solution of Eqs. (3.39) and (3.40) is also a solution of the system of equations (3.43).

Quadratic Lagrangians

The variation of the components of the energy-momentum tensor leads to an additional set of 16 linear equations (without assuming the symmetry of the energy-momentum tensor):

$$\frac{\kappa}{2} \varepsilon^{\rho\sigma\delta[\mu} g^{\nu]\beta} C^E_{\mu\nu\rho\sigma} + \frac{\kappa}{3} \varepsilon^{\mu\nu\rho\sigma} C^E_{\mu\nu\rho\sigma} g^{\beta\delta} = 0 \ . \tag{3.44}$$

Obviously, any solution of Eq. (3.42) is also a solution of Eq. (3.44).

To summarize, we obtained a system of 112 linear algebraic equations in 112 field variables of the tensors C and T, and as a result the gravitational and energy-momentum tensors are expressed in terms of the metric tensor and its derivatives. Furthermore, any solution of the Einstein field equations is also a solution of the system of algebraic equations.

3.6.4 The variation of the metric tensor

In the last subsection we obtained a linear algebraic system, where the number of equations is equal to the number of unknown variables. In this subsection the variation of the metric tensor is shown to imply an additional set of 10 quadratic algebraic equations relating the components of the same tensors, these equations become identities when the Einstein field equations are taken into account. The detailed calculation will be omitted since, if in the Lagrangian (3.30) one substitutes Eqs. (3.39) and (3.40), the scalar density of Lanczos emerges, generating identically vanishing Lagrange equations. Thus the solutions of Eqs. (3.39) and (3.40) satisfy the Lagrange equations obtained from the variations of the metric tensor.

In the following we show that by adding the scalar density $\mathcal{L}_6$, given by Eq. (3.36), to the total Lagrangian (3.30), it is possible to let the variables $g^{\alpha\beta}$ and $\Gamma^\alpha_{\beta\alpha}$ vary independently, and this has some ramifications on the search for conservation laws. Under these circumstances the only derivatives remaining in the Lagrangian are the first derivatives of the Γ's. Since the metric tensor appears in the total Lagrangian and in the scalar density $\mathcal{L}_6$ both explicitly and through the Christoffel symbols, the variation of $\mathcal{L} + \mathcal{L}_6$ induced by the variation of the metric tensor is given by

$$\delta_g \bar{\mathcal{L}} = \frac{\delta \bar{\mathcal{L}}}{\delta \Gamma^\alpha_{\beta\gamma}} \delta_g \Gamma^\alpha_{\beta\gamma} + \left[\frac{\partial \bar{\mathcal{L}}}{\partial g_{\alpha\beta}} \right]_\Gamma \delta g_{\alpha\beta} + \text{divergence} \ , \tag{3.45}$$

where

$$\bar{\mathscr{L}} = \mathscr{L} + \mathscr{L}_6 ,$$

and

$$\frac{\delta \bar{\mathscr{L}}}{\delta \Gamma^\alpha_{\beta\gamma}} = \frac{\partial \bar{\mathscr{L}}}{\partial \Gamma^\alpha_{\beta\gamma}} - \frac{\partial}{\partial x^\delta} \left[\frac{\partial \bar{\mathscr{L}}}{\partial \Gamma^\alpha_{\beta\gamma,\delta}} \right] .$$

Assuming the symmetry $\Gamma^\mu{}_{\alpha\beta} = \Gamma^\mu{}_{\beta\alpha}$, we find

$$\frac{\delta \bar{\mathscr{L}}}{\delta \Gamma^\beta_{\nu\sigma}} = -\varepsilon^{\gamma\delta\rho\sigma}(C^\nu{}_{\beta\gamma\delta} + \kappa U^\nu{}_{\beta\gamma\delta})_{;\rho} . \qquad (3.46)$$

From this, together with the Bianchi identities (3.20), it follows that the first term in Eq. (3.45) vanishes for values of $C^\alpha{}_{\beta\gamma\delta}$ and $T^{\alpha\beta}$ which satisfy the Einstein field equations.

A straightforward, though tedious, calculation shows that the second term in Eq. (3.45) also vanishes for the values of the variables that satisfy Einstein's field equations. According to what has been said at the beginning of this section, this calculation will be omitted.

3.7 Summary

In this chapter we have discussed the action principle for the Einstein gravitational field equations. Our starting point was Hilbert's celebrated linear Lagrangian, but the main body of the chapter has been devoted to quadratic Lagrangians leading to the Einstein field equations. The quadratic Lagrangian approach could be valuable in relation to the quantization program of the gravitational field, due to the fact that gravity based on quadratic Lagrangian becomes more similar to the ordinary SU(2) gauge field theory, which has been shown by 't Hooft to be renormalizable.

We gave a review of the quadratic invariants constructed out of the Riemann tensor, two of which, the Lanczos invariants, are related to the quadratic Lagrangians discussed above.

We have examined Carmeli's quadratic Lagrangian which produces the Newman-Penrose version of the Einstein field equations. The existence of a one-parameter family of Lagrangians has been demonstrated, one member of which is invariant under the group SL(2,C), and is equal, in its extremal value, to Carmeli's Lagrangian.

A tensorial quadratic Lagrangian, due to Nissani, which gives the Einstein gravitational field equations of general relativity in their tensorial form, has been exhibited. This Lagrangian, which is a scalar density, is a function of three field variables: the metric tensor, the energy-momentum tensor, and the gravitational tensor. The first two tensors have *a priori* clear physical meaning, whereas the determination of the third tensor as the Weyl tensor is a consequence of the requirement for an extremal value for the action.

This last Lagrangian, in the extremal value of the action, is equal to the spurious Lanczos invariant. By replacing Lanczos' variational method, which leads to identically vanishing Lagrange expressions, by a Palatini-type method, we arrive at the full Einstein gravitational field equations. It is worthwhile mentioning that there are no absolute variables (on which no variation is carried out) in the Lagrangian, and there is no need to resort to a spinor space.

Finally we mention that the tensorial quadratic Lagrangian is a sum of a gravitational term and a term corresponding to an interaction between gravitation and matter. One could have added a third term describing matter. Such a term, however, has no effect on the stationary value of the action.

4
ENERGY CONSERVATION PROBLEM IN GENERAL RELATIVITY

4.1 *Introduction*

The problem of energy conservation has plagued the theory of general relativity since its very inception[51–53]. As is well-known, the conservation of energy-momentum is expressed, in special relativity theory, by the vanishing of the ordinary divergence of the energy-momentum tensor. Passing from the Minkowskian flat space-time to the pseudo-Riemannian curved space-time, this ordinary divergencelessness is replaced, according to the general covariance principle with "minimal coupling", by covariant divergencelessness. The source of the problem lies in the fact that unlike ordinary divergencelessness, the vanishing of the covariant divergence does not, in general, constitute a continuity equation, and consequently is not a manifestation of a conservation law in the usual sense. Thus, in general relativity theory the energy-momentum tensor is conserved only in those coordinate systems, and under the conditions, in which the two divergences of the energy-momentum tensor—the ordinary and the covariant—coincide.

Indeed, the covariant divergence at a point A does reduce to the ordinary divergence in a coordinate system which is geodesic at this point. Since laboratory verifications of the energy-momentum conservation (excluding gravitational energy) invariably employ geodesic systems, the compatibility of the experimental results with the theoretical proposition has been taken as a corroboration of the latter. On the other hand, the local strict conservation, together with the global nonconservation, runs against the principle of the localized nature of physical processes. It appears as if general relativity dictates that energy-momentum is created or annihilated as a global process over finite domain of space-time, even though at each point of that domain strict conservation prevails!

Throughout the seven decades since the publication of the solution offered by Einstein himself, many approaches to the problem have been proposed, and there is a vast literature devoted to this subject. The scope of this book allows us to cover only a portion of this wealth of material. It is generally accepted that all the proposed solutions suffer from some difficulties, and none can be considered as a fully satisfactory resolution[54-56]. We close this chapter with a discussion of the conserved tensor density found by Nissani[26] in the framework of the SL(2,C) gauge theory of gravitation[22], which is distinguished among the proposed expressions for the total energy-momentum, material plus gravitational, in its being tensorial.

4.2 The Problem

In special relativity theory the energy-momentum conservation law is expressed by the ordinary divergencelessness of the energy-momentum tensor, i.e., by the continuity equation:

$$T^{\alpha\beta}{}_{,\beta} = 0 . \qquad (4.1)$$

According to the principle of equivalence, this equation ought to remain valid, in curved space-time, at least locally in geodesic coordinate systems[57,58]. On the other hand, the principle of general covariance dictates that the conservation law should be expressible in a form valid in all coordinate systems. These two requirements are simultaneously satisfied by requiring the covariant divergencelessness condition of the energy-momentum tensor,

$$T^{\alpha\beta}{}_{;\beta} = T^{\alpha\beta}{}_{,\beta} + \Gamma^{\beta}_{\beta\rho}T^{\alpha\rho} + \Gamma^{\alpha}_{\beta\rho}T^{\rho\beta} = 0 , \qquad (4.2)$$

Energy Conservation Problem in General Relativity

since, due to the vanishing of the Christoffel symbols in the geodesic systems of coordinates, Eq. (4.2) reduces to (4.1) in these systems.

Equation (4.2) itself, in turn, appears in general relativity as a consequence of Einstein's field equations

$$G^{\mu\nu} = \kappa T^{\mu\nu}, \tag{4.3}$$

and the identically general-covariant divergencelessness of Einstein's tensor (contracted Bianchi identity)

$$G^{\mu\nu}{}_{;\nu} = 0, \tag{4.4}$$

where

$$G^{\mu\nu} = R^{\mu\nu} - \frac{1}{2} g^{\mu\nu} R$$

is the Einstein tensor, and

$$R^{\mu\nu} = R^{\rho\mu}{}_\rho{}^\nu, \qquad R = R^\mu{}_\mu$$

are the Ricci tensor and scalar, respectively.

Now, using the identity

$$\Gamma^\mu_{\rho\mu} = \frac{1}{\sqrt{-g}} \frac{\partial \sqrt{-g}}{\partial x^\rho}, \tag{4.5}$$

Eq. (4.2) can be cast in the form

$$\sqrt{-g} T^{\alpha\beta}{}_{;\beta} = (\sqrt{-g} T^{\alpha\beta})_{,\beta} + \sqrt{-g} \Gamma^\alpha_{\beta\rho} T^{\beta\rho} = 0, \tag{4.6}$$

or equivalently (for future reference),

$$\sqrt{-g} T_\alpha{}^\beta{}_{;\beta} = (\sqrt{-g} T_\alpha{}^\beta)_{,\beta} - \sqrt{-g} \Gamma^\rho_{\alpha\beta} T_\rho{}^\beta = 0. \tag{4.7}$$

The term involving Γ reduces to

$$\sqrt{-g}\Gamma^{\rho}_{\alpha\beta}T_{\rho}{}^{\beta} = \frac{1}{2}\sqrt{-g}g^{\rho\sigma}(g_{\alpha\sigma,\beta} + g_{\beta\sigma,\alpha} - g_{\alpha\beta,\sigma})T_{\rho}{}^{\beta}$$

$$= \frac{1}{2}\sqrt{-g}g_{\sigma\beta,\alpha}T^{\sigma\beta} , \qquad (4.8)$$

and hence Eq. (4.7) assumes the form

$$(\sqrt{-g}T_{\alpha}{}^{\beta})_{,\beta} = \frac{1}{2}\sqrt{-g}g_{\sigma\beta,\alpha}T^{\sigma\beta} . \qquad (4.9)$$

This version will be useful in the next section.

Equation (4.6) differs from a continuity equation by the presence of the Γ term. The latter resembles the force density, which is, in special relativity theory, equal to the divergence of a partial energy-momentum tensor. Thus, if $S_{\alpha}{}^{\beta}$ is the electromagnetic stress-energy tensor of a special relativistic system, its divergence satisfies

$$S_{\alpha}{}^{\beta}{}_{,\beta} = f_{\alpha} = f_{\alpha}{}^{\beta}j_{\beta} ,$$

where the 4-force f_{α} is given by the contraction of the Maxwell field tensor $f_{\alpha}{}^{\beta}$ with the current 4-vector j_{β}. Notice the form of this force—a product of the field f taken into account in the stress-energy tensor S, and the current j, which reflects the motion of the charge sources not included in S. The general relativistic situation embodied in Eq. (4.6) is analogous, with one exception. While the term expressing the force density in flat space-time has its real significance in inertial coordinate systems, the term we are dealing with vanishes precisely in the geodesic systems, which play the role of the "inertial" systems in curved space-time.

The right-hand side in Eq. (4.9) has been interpreted by Einstein and his followers, in analogy with the flat space-time case, as expressing the interaction between two kinds of energy: the matter energy, which is included in the tensor T appearing on the right-hand side of the Einstein field equations, and the gravitational energy, which is not included in T. This interpretation is suggested by the particular form of this term, namely, a product of a metric-dependent quantity (expressing gravitation) and T (expressing the matter energy). The fact that the right-hand side of Eq. (4.9) admits the interpretation of an interaction originating from gravitational energy, gave rise to the idea that finding a mathematical expression for the gravitational energy, and adding it to the energy-momentum tensor, would lead to the continuity equation

Energy Conservation Problem in General Relativity

$$(\sqrt{-g}T^{\mu\nu} + \sqrt{-g}t^{\mu\nu})_{,\nu} = 0 . \qquad (4.10)$$

This equation has to be valid in all coordinate systems, namely, to lead to a general-covariant conservation law (in the sense of a law holding in all coordinate systems).

The vanishing of the ordinary divergence of a symmetric tensor density is not a general covariant condition. Hence, requiring that (4.10) be a general covariant condition is inconsistent with $t^{\mu\nu}$ being both symmetric and tensorial. The latter are two very desirable features for an energy-momentum expression to possess. Lack of symmetry in $t^{\mu\rho}$ entails difficulties in defining angular momentum. The nontensorial nature of $t^{\mu\nu}$, on the other hand, makes the physical interpretation of the gravitational energy quite problematic: boundary conditions and other constraints must be imposed, and they in turn restrict the class of admissible coordinate systems, thereby violating the principle of general covariance[54,59–63]. (For a discussion of possible tensorial but nonsymmetric expression of the total energy-momentum, see Sec. 4.6.)

4.3 Einstein's Solution

As pointed out in the previous section, the Einstein school of thought concerning the problem of energy conservation in general relativity adheres to the following guidelines:

(a) Search for the expression $t^{\alpha\beta}$ of the gravitational energy-momentum density.

(b) This gravitational energy-momentum density should complement the matter energy-momentum density to an ordinary divergenceless expression. Thus, in conjunction with $T^{\alpha\beta}$ it should constitute the basis of a continuity equation.

(c) This continuity equation is to be valid in all coordinate systems, namely, it should express a general-covariant conservation law.

In order to realize this scheme, the following mathematical observation proves to be useful. Suppose that a physical theory of the fields Φ is derived from the Lagrangian $\mathscr{L} = \mathscr{L}(\Phi, \Phi_{,\alpha}, x)$. Hence the field equations governing the dynamics of the theory are

$$\frac{\partial}{\partial x^\mu} \frac{\partial \mathscr{L}}{\partial \Phi_{,\mu}} - \frac{\partial \mathscr{L}}{\partial \Phi} = 0 . \qquad (4.11)$$

This Lagrange equation implies

$$\left[\frac{\partial \mathscr{L}}{\partial \Phi_{,\nu}}\Phi_{,\mu} - \delta_\mu^\nu \mathscr{L}\right]_{,\nu} = (\sqrt{-g}\tau_\mu{}^\nu)_{,\nu} = -\frac{\partial \mathscr{L}}{\partial x^\mu}, \qquad (4.12)$$

where the right-hand side reflects the explicit dependence of the Lagrangian on the coordinates. Consequently, if the Lagrangian does not depend explicitly on the coordinates, Eq. (4.12) reduces to the divergencelessness condition,

$$(\sqrt{-g}\tau_\mu{}^\nu)_{,\nu} = 0, \qquad (4.13)$$

valid in all coordinate systems, i.e., we have at our disposal a general-covariant conservation law (general-covariant in the sense of a law holding in all coordinate systems). Note, however, that $\tau_\mu{}^\nu$ does not have tensorial character.

The question, then, arises as to which Lagrangian should be taken to produce the conserved $\tau_\mu{}^\nu$. To employ the scheme outlined above, it ought to depend on the field variables and their first-order partial derivatives alone, and of course generate the field equations as its Lagrange equations. Einstein, in his original work[1], observed that the Lagrangian

$$\mathscr{L}_E = g^{\mu\nu}\Gamma^\alpha_{\mu\beta}\Gamma^\beta_{\nu\alpha}, \qquad (4.14)$$

considered as a function of the metric tensor and its first derivatives, with the additional constraint

$$\sqrt{-g} = 1, \qquad (4.15)$$

will do.

Obviously, the condition (4.15) restricts the allowed coordinate systems, which is not consistent with the spirit of the principle of general covariance. This shortcoming, however, can be easily removed by noting that on account of (4.15), the Lagrangian (4.14) may be written in the form

$$\mathscr{L}_E = \sqrt{-g}\,g^{\mu\nu}(\Gamma^\sigma_{\mu\rho}\Gamma^\rho_{\nu\sigma} - \Gamma^\sigma_{\mu\nu}\Gamma^\rho_{\rho\sigma}), \qquad (4.16)$$

the second term vanishing since,

$$\Gamma^\beta_{\beta\alpha} = \frac{1}{\sqrt{-g}}\frac{\partial \sqrt{-g}}{\partial x^\alpha} = 0. \qquad (4.17)$$

Energy Conservation Problem in General Relativity

The Lagrangian (4.16) in any coordinate system (i.e., without the auxiliary condition (4.17)), generates, through a variational principle, the field equations of general relativity, and is dependent on the metric tensor components and their first-order derivatives alone. As is now customary, we will consider a variation of Einstein's idea, namely, our basic Lagrangian will be (4.16). The latter, in fact, is identical with the Hilbert Lagrangian, up to a divergence term, viz.,

$$\mathcal{L}_H = \sqrt{-g}R = \mathcal{L}_E + [\sqrt{-g}(g^{\alpha\mu}\Gamma^\rho_{\mu\rho} - g^{\mu\nu}\Gamma^\alpha_{\mu\nu})]_{,\alpha} \ . \tag{4.18}$$

The last identity implies that the two Lagrangians are equivalent, i.e., both lead to the same field equations (the vacuum Einstein equations). While $\mathcal{L}_E$ has the deficiency of being nontensorial (unlike $\mathcal{L}_H$), it has the advantage of depending only on the metric tensor components and their first-order derivatives, rendering it suitable for the application of the algorithm (4.12), to yield

$$\kappa(\sqrt{-g}\tau^\nu_{E\mu})_{,\nu} = \left[\frac{\partial \mathcal{L}_E}{\partial g_{\alpha\beta,\nu}}g_{\alpha\beta,\mu} - \delta^\nu_\mu \mathcal{L}_E\right]_{,\nu} = 0 \ . \tag{4.19}$$

The nontensorial divergenceless complex $\sqrt{-g}\tau^\nu_{E\mu}$ of 4×4 physical quantities is derived from the gravitational Lagrangian through the same process which is the case of other fields (e.g. the electromagnetic field) produces the energy-momentum tensor. The Einstein approach outlined here, therefore, interprets this complex as the expression of the energy-momentum of the gravitational field.

It is worth noticing that while $\mathcal{L}_E$ is not a scalar density with respect to general coordinate transformations, it is a scalar density with respect to affine transformations, as follows from the fact that with respect to affine transformations the connections behave as tensors. Likewise, since the partial derivatives of the metric tensors are tensorial with respect to the affine transformations, the quantities $\tau_E{}^\nu{}_\mu$ are tensorial with respect to the affine group.

So far we have dealt with empty space. In the presence of matter, as will be now shown, one finds that the sum

$$\tau^\nu_{E\mu} = \sqrt{-g}T^\nu{}_\mu + \sqrt{-g}\tau^\nu_{E\mu} \ ,$$

where $T^\nu{}_\mu$ is the matter energy-momentum tensor, is conserved:

$$\tau^\nu_{E\mu,\nu} = (\sqrt{-g}T^\nu{}_\mu + \sqrt{-g}\tau^\nu_{E\mu})_{,\nu} = 0 \ . \tag{4.20}$$

This provides further support to the identification of $\tau_E{}^\nu{}_\mu$ as the expression of the energy-momentum of the gravitational field. To prove (4.20), consider the formula (4.12) in the special case where the fields Φ are the metric field $g_{\alpha\beta}$ and additional physical fields ϕ, and the Lagrangian can be decomposed:

$$\mathcal{L} = \mathcal{L}_E + \mathcal{L}_M(g_{\alpha\beta}, \phi, \phi_{,\beta}) . \qquad (4.21)$$

Here it is assumed that the matter Lagrangian $\mathcal{L}_M$ does not depend on the derivatives of the metric. Under these assumptions Eq. (4.12) yields:

$$\left[\frac{\partial \mathcal{L}_E}{\partial g_{\alpha\beta,\nu}} g_{\alpha\beta,\mu} + \frac{\partial \mathcal{L}_M}{\partial \phi_{,\nu}} \phi_{,\mu} - \delta^\nu_\mu \mathcal{L}_E - \delta^\nu_\mu \mathcal{L}_M \right]_{,\nu} = 0 , \qquad (4.22)$$

or, using Eq. (4.19),

$$\kappa (\sqrt{-g}\tau_{E\mu}^\nu)_{,\nu} + \left[\frac{\partial \mathcal{L}_M}{\partial \phi_{,\nu}} \phi_{,\mu} \right]_{,\nu} - \mathcal{L}_{M,\mu} = 0 . \qquad (4.23)$$

Taking into account the Lagrange equations corresponding to the variation of the fields ϕ, the last equation reduces to

$$\kappa(\sqrt{-g}\tau_{E\mu}^\nu)_{,\nu} - \frac{\partial \mathcal{L}_M}{\partial g_{\alpha\beta}} g_{\alpha\beta,\mu} = 0. \qquad (4.24)$$

The last term in this equation, by the definition of the matter energy-momentum tensor, is equal to $\kappa\sqrt{-g}T^{\alpha\beta}g_{\alpha\beta,\mu}$. Hence, in view of (4.9), the last equation is equivalent to (4.20).

The Einstein pseudo-tensor τ_E, in addition to lacking tensorial character (like all the other expressions proposed for the gravitational energy-momentum), suffers from two major drawbacks:

(a) $\tau_E{}^{\mu\nu}$ is not symmetric in the indices μ and ν. Thus, no definition of conserved angular momentum can be based upon it.

(b) It depends on the metric components and their first derivatives only, and it vanishes at the origin of geodesic coordinate systems, which means that locally the gravitational energy-momentum is absent under the normal circumstances of laboratory measurements. This casts grave doubts on the physical significance of this pseudo-tensor. It is customary to interpret this anomaly as the manifestation of a peculiarity of the gravitational energy, namely, that unlike matter energy, the gravitational energy is global in its nature, and cannot be localized in space. (The meaning of the term "global" in this context differs from its meaning in the rest of this book.)

Consider a physical system bounded within a 3-volume V. In special relativity the integral

$$P^\alpha = \int_V T^{\alpha 0} d^3x \qquad (4.25)$$

of the components $T^{\alpha 0}$ of the energy-momentum tensor density over the volume V is the energy-momentum 4-vector of the system. In general relativity, however, an expression of the form (4.25)—an integral of tensor components—does not possess tensorial character, since under a general coordinate change the transformation coefficients are functions of the coordinates. Only by restricting the allowed transformations to the affine group one can assign tensorial nature to an integral of a tensor density. It has been remarked already that under the affine group τ_E transforms as a tensor. It follows that τ_E is an affine tensor density of weight $+1$. Consequently, the integral

$$P_\alpha = \int_V \sqrt{-g}\, \tau^0_{E\alpha} d^3x \qquad (4.26)$$

defines an affine vector, which, under certain conditions on the coordinate system, is conserved and represents the energy-momentum of the system.

The problem associated with the lack of symmetry of the Einstein pseudo-tensor [(a) above] are remedied by the solution proposed by Landau and Lifshitz, who constructed a symmetric pseudo-tensor, which will be discussed in the next section. The problem of the nonlocalizability of the Einstein gravitational energy [(b) above] is tackled by Moller, whose proposed solution will be discussed in Sec. 4.5.

4.4 *Landau-Lifshitz Pseudo-Tensor*

In the previous section the Einstein pseudo-tensor was discussed. It was pointed out that it is not symmetric in the two indices, and as a consequence it cannot serve as a basis for the definition of conserved angular momentum. This constitutes a serious deficiency of the Einstein pseudo-tensor, since the inability to define gravitational angular momentum is not less disturbing than the difficulties originating with the apparent nonconservation of the energy-momentum. To overcome this problem, Landau and Lifshitz[64] postulated an alternative pseudo-tensor, which, unlike the expression proposed by Einstein, is symmetric. Their pseudo-tensor too has the property that by adding it to the matter energy-momentum tensor density one obtains a physical quantity which is conserved in any coordinate system. Furthermore, with the aid of this pseudo-tensor, conserved angular momentum can be defined in the usual manner.

To derive the expression of the symmetric pseudo-tensor, consider the Einstein field equations at the origin of a geodesic coordinate system, where the connections vanish. The surviving part of the Einstein tensor is

$$G^{\mu\nu} \doteq \frac{\partial}{\partial x^\rho}\left[\frac{\partial}{\partial x^\sigma}[(-g)(g^{\mu\nu}g^{\rho\sigma} - g^{\mu\rho}g^{\nu\sigma})]\right], \qquad (4.27)$$

where the dot above the equality sign indicates that this equality holds at the origin of a geodesic system, but not in general coordinate systems. Notice that the right-hand side of this equation has the form of a partial derivative with respect to x^ρ of an expression which is antisymmetric in the indices ρ and ν. Consequently, differentiating it once more with respect to x^ν yields zero. Thus if the expression on the right-hand side, *in any coordinate system*, is denoted by

$$(-g)\tau_L^{\mu\nu} = \frac{1}{\kappa}\frac{\partial}{\partial x^\rho}\left[\frac{\partial}{\partial x^\sigma}[(-g)(g^{\mu\nu}g^{\rho\sigma} - g^{\mu\rho}g^{\nu\sigma})]\right], \qquad (4.28)$$

then $(-g)\tau_L^{\mu\nu}$ is a symmetric quantity with vanishing ordinary divergence, and the Einstein equations reduce, in geodesic systems, to

$$\tau_L^{\mu\nu} \doteq T^{\mu\nu}.$$

Clearly, the two sides of the last equation behave differently under a coordinate transformation, hence they will not be equal to each other in general coordinate systems. Their difference is defined by Landau and Lifshitz to be the gravitational energy-momentum pseudo-tensor:

$$t_L^{\mu\nu} = \tau_L^{\mu\nu} - T^{\mu\nu}.$$

Indeed, adding this pseudo-tensor to the matter energy-momentum tensor, one gets,

Energy Conservation Problem in General Relativity 55

$$[(-g)\tau_L^{\mu\nu}]_{,\nu} = [(-g)(T^{\mu\nu} + t_L^{\mu\nu})]_{,\nu} = 0 \ . \tag{4.29}$$

Let us now assess the merits of the Landau and Lifshitz pseudo-tensor, compared with the Einstein pseudo-tensor. Both of them have the undesired property of being nontensorial, and they vanish at the origin of geodesic systems. Both behave as tensors under the group of affine transformations. This implies that the Landau-Lifshitz conserved quantity is a tensor density of weight +2 with respect to the affine group, and hence

$$(-g)\tau_L^{\alpha 0} d^3x = dP^\alpha \tag{4.30}$$

is not an affine vector, but rather an affine vector density of weight +1. This stands in contrast to the situation in the case of the Einstein pseudo-tensor, where the integral over spatial volume yields a vector with respect to the affine group. Here, on the other hand, the integral represents a vector only with respect to a more restricted group—the linear transformations with unit determinant (e.g., Lorentz transformations).

An advantage of $\tau_L^{\mu\nu}$ over $\tau_E^{\mu}{}_\nu$ is its being symmetric in the two indices, a fact which allows the definition of a conserved angular-momentum complex

$$M^{\alpha\beta\gamma} = (-g)(x^\alpha \tau_L^{\beta\gamma} - x^\beta \tau_L^{\alpha\gamma}) \ . \tag{4.31}$$

It follows immediately from the symmetry and divergencelessness of the total energy-momentum pseudo-tensor that the angular-momentum complex as defined above is conserved:

$$M^{\alpha\beta\gamma}{}_{,\gamma} = 0 \ . \tag{4.32}$$

It is to be noted that the Einstein and Landau-Lifshitz expressions for the total energy (P_0 and P^0) share the undesirable trait that they are not invariant even under the restricted class of coordinate transformations of the form

$$x'^0 = x^0$$

$$x'^j = f^j(x^i) \ .$$

Thus, the total energy at time x^0 depends on the chart used to label points on the hypersurface $x^0 = $ const.. For example, in flat space one finds for the Einstein expression of the total energy:

$$P_0 = 0 \text{ in Cartesian coordinates}$$

and

$$P_0 = -\infty \text{ in spherical coordinates,}$$

as was pointed out by Bauer[65].

The approach outlined in the next section overcomes this difficulty.

4.5 *Moller's Approach*

An insight into the nature of the pseudo-tensors is gained by observing that the conserved total energy-momentum according to Landau and Lifshitz (see Sec. 4.4) is the ordinary divergence of a 3-index complex, called "superpotential", which is antisymmetric in the last two indices:

$$(-g)\tau_L^{\mu\nu} = \Phi_L^{\mu\nu\rho}{}_{,\rho}, \qquad \Phi_L^{\mu\nu\rho} = -\Phi_L^{\mu\rho\nu}, \tag{4.33}$$

with

$$\Phi_L^{\mu\nu\rho} = \frac{1}{\kappa}[(-g)(g^{\mu\nu}g^{\rho\sigma} - g^{\mu\rho}g^{\nu\sigma})]_{,\sigma}. \tag{4.34}$$

In fact, let

$$\Phi^{\mu\nu\rho} = -\Phi^{\mu\rho\nu}$$

be *any* antisymmetric "superpotential". Then, if one defines

$$\tau^{\mu\nu} = \Phi^{\mu\nu\rho}{}_{,\rho},$$

it clearly satisfies identically (i.e., independent of the particular structure of the "superpotential")

$$\tau^{\mu\nu}{}_{,\nu} = 0.$$

This procedure is a general algorithm to generate conserved quantities, of which the Landau-Lifshitz complex is a special case corresponding to a particular choice of the superpotential. Moreover, it has been shown by von Freud[66] that the Einstein conserved total energy-momentum pseudo-tensor can be constructed by the same algorithm, namely, by choosing the superpotential

$$\Phi_{E\mu}^{\nu\rho} = \frac{g_{\mu\alpha}}{2\kappa\sqrt{-g}} \frac{\partial}{\partial x^\sigma}[(-g)(g^{\alpha\nu}g^{\rho\sigma} - g^{\alpha\rho}g^{\nu\sigma})] \ . \tag{4.35}$$

Then the Einstein total energy-momentum complex is given by

$$\sqrt{-g}\,\tau_{E\mu}^{\nu} = \Phi_{E\mu}^{\nu\rho}{}_{,\rho} \ . \tag{4.36}$$

The fact that the conserved total energy-momentum pseudo-tensor has a representation as the ordinary divergence of a superpotential entails a curious mathematical result with interesting physical significance. According to the last section, the total energy-momentum content of a physical system bounded in the 3-volume V is given by [Eq. (4.26)]

$$P_\mu = -\int_V \tau_\mu^0 d^3x \ . \tag{4.37}$$

If the total energy-momentum pseudo-tensor is derived from a superpotential, then

$$P_\mu = -\int_V \Phi_\mu^{0\rho}{}_{,\rho} d^3x \ .$$

Due to the antisymmetry of the superpotentials, in the equation above the summation is carried out only over the three spatial values of the index ρ, and by applying the Gauss theorem in three dimensions we find:

$$P_\mu = -\int_\Sigma \Phi_\mu^{0j} n_j dS \ , \tag{4.38}$$

where Σ is the two-dimensional boundary of the volume V, and n_j is the unit vector normal to the surface element dS of Σ. Hence the total energy-momentum of the system is determined solely by the value of the metric and its derivatives on the *surface* of the volume V, the specific form of the metric inside the volume being immaterial.

To remedy some of the difficulties exhibited by the Einstein and Landau-Lifshitz complexes, Moller[67,68] suggested using the superpotential

$$\Phi_{M\mu}^{\nu\gamma} = \frac{\sqrt{-g}}{\kappa}(g_{\mu\rho,\sigma} - g_{\mu\sigma,\rho})g^{\nu\sigma}g^{\tau\gamma} \ , \tag{4.39}$$

leading to the total and gravitational energy-momentum pseudo-tensors defined by

$$\Phi^{\nu}_{M\mu}{}^{\gamma}{}_{,\gamma} \equiv \sqrt{-g}\tau^{\nu}_{M\mu} \equiv \sqrt{-g}(T^{\nu}{}_{\mu} + \tau^{\nu}_{M\mu}) \ . \tag{4.40}$$

The total energy-momentum pseudo-tensor satisfies the continuity equation

$$(\sqrt{-g}\tau^{\nu}_{M\mu})_{,\nu} = 0 \ , \tag{4.41}$$

thus providing another general-covariant conservation law. It follows from Eq. (4.39) that

$$\sqrt{-g}\tau^{0}_{M0}d^{3}x = dP^{0}_{M} \tag{4.42}$$

is a scalar with respect to spatial coordinate transformations, hence the total energy has a well-defined value for the hypersurface $x^0 = $ const., independent of the chart used on the hypersurface. Thus, the Bauer objection does not apply here. This is an advantage of the Moller complex over the Einstein and Landau-Lifshitz complexes.

The Moller complex is also free from another problem plaguing the other two complexes, namely, the nonlocalizabilty of the energy-momentum. Since the Moller pseudo-tensor depends on the second derivatives of the metric, it cannot be eliminated at an arbitrary point by using a geodesic coordinate system. The Moller complex, however, compared with the other two complexes, has the disadvantage of not being an affine tensor.

We see that none of the expressions for the gravitational energy discussed so far can be considered trouble-free. Moller concludes[61,62] that in order to construct an expression which will satisfy a minimal set of physical requirements, the metric alone is not sufficient, and more elements must be incorporated[a]. Those new ingredients might be of the form of an auxiliary flat metric, as in the Rosen bi-metric theory[69,70], or a tetrad field[56,71,72], which will now be briefly discussed. Assign to each point of space-time a set of four independent vectors ("tetrad") $h_{A\mu}$, where the tetrad index A runs from 0 to 3. The tetrad field is tied up with the metric by the relation

$$g_{\mu\nu} = h^{A}_{\mu}h_{A\nu} \ , \tag{4.43}$$

---

[a] We would like to thank S. Malin for drawing our attention to the relation between Moller's conclusion and the conserved tensor density[26] discussed in the next section.

where tetrad indices are raised and lowered using the Minkowskian metric. In terms of the tetrad field, the gravitational Lagrangian can be cast in the form

$$\mathscr{L}_M = \sqrt{-g}(\theta_{\mu\nu\rho}\theta^{\rho\nu\mu} - \phi^\rho\phi_\rho) , \qquad (4.44)$$

where

$$\theta_{\mu\nu\rho} = h^A_\mu h_{A\nu;\rho} ; \qquad \phi_\rho = \theta^\mu{}_{\rho\mu} , \qquad (4.45)$$

(with semicolon denoting covariant derivatives).

Since θ and ϕ are tensors, this Lagrangian, which is first order in the variables (the tetrad components), is a scalar density with respect to the group of general coordinate transformations. (Recall that the Einstein Lagrangian, which is also first-order, is a tensor density only with respect to the affine group.) The Moller tetrad Lagrangian differs from the Hilbert Lagrangian by a divergence term. From this Lagrangian one derives, through the usual procedure, Eq. (4.19), the expression for the gravitational energy-momentum

$$\bar{\tau}^\nu_{M\mu} = \frac{1}{2\kappa}\left[\frac{\partial \mathscr{L}_M}{\partial h^A_{\rho,\nu}}h^A_{\rho,\mu} - \delta^\nu_\mu \mathscr{L}_M\right] . \qquad (4.46)$$

The total energy-momentum complex is generated by the antisymmetric superpotential

$$\bar{\Phi}^{\nu}_{M\mu}{}^\rho = \frac{\sqrt{-g}}{\kappa}(\theta^{\nu\rho}{}_\mu - \delta^\nu_\mu \phi^\rho + \delta^\rho_\mu \phi^\nu) . \qquad (4.47)$$

We have, therefore,

$$\bar{\Phi}^{\nu}_{M\mu}{}^\rho{}_{,\rho} \equiv \sqrt{-g}\bar{\mathcal{T}}^\nu_{M\mu} = \sqrt{-g}(T^\nu{}_\mu + \bar{\tau}^\nu_{M\mu}) , \qquad (4.48)$$

and

$$(\sqrt{-g}\bar{\mathcal{T}}^\nu_{M\mu})_{,\nu} = 0 .$$

To assess the merits of this energy-momentum complex, we observe that it too, like the Einstein and Landau-Lifshitz complexes, is an affine tensor. Furthermore, it behaves well under spatial coordinate transformations, and it is not subject to the Bauer criticism. On the other hand, the gravitational energy is not determined by the metric tensor, since it is defined in terms of the tetrad components, and they, in turn, are not determined uniquely by the metric. Finally, as in the cases discussed previously, the nontensorial nature of the complex implies that only by utilizing boundary conditions and special coordinate systems one can assign to it physical meaning.

In the next section another expression for the conserved energy-momentum will be introduced from the standpoint of the SL(2,C) gauge theory formulation of general relativity. The expression obtained thereby does not exhibit some of the problems pointed out for the expressions which have been discussed here so far.

4.6 Conserved Tensor Density—Gauge Formulation Approach

The problem encountered by the various attempts to define conserved total energy-momentum had led Moller, as mentioned in the previous section, to the conclusion that the metric alone cannot be the basis for a satisfactory definition of the energy-momentum, and some external fields must be infused into the formulation. Two possible candidates for the additional ingredients were mentioned in the last section: the auxiliary flat metric of the Rosen bi-metric theory, and the tetrad structure of the Moller approach. In this section we will show, following Nissani[26], that the needed additional elements can be found in the spinor space, tangent to the pseudo-Riemannian manifold, in the framework of the SL(2,C) gauge theory formulation[2-22] of the Einstein theory of gravitation.

As was shown in Chapter 2, the gauge group of the Carmeli gauge theory is SL(2,C), and it acts on a two-dimensional complex linear space attached to each point of the 4-dimensional pseudo-Riemannian space-time manifold. The gauge field is defined by

$$F_{\mu\nu A}{}^B = B_{\mu A}{}^B{}_{,\nu} - B_{\nu A}{}^B{}_{,\mu} + [B_\mu, B_\nu]_A{}^B . \qquad (4.49)$$

Here B and F are tensors of order one and two respectively, the components of which are 2×2 matrices with complex entries. The Latin indices A, B are spinor indices, and run through the values 0 and 1. Under the action of the group of gauge transformations F transforms as a 2-spinor, while the vector potential B undergoes a nonhomogeneous transformation.

The field F is the spinorial expression of the curvature of the manifold, and is connected to the Riemann curvature tensor R by the relation

$$F_{\mu\nu A}{}^B = \frac{1}{2} R^\rho{}_{\sigma\mu\nu} \sigma_{\rho AC'} \sigma^{\sigma BC'} . \tag{4.50}$$

In the last equation, the connecting quantities σ constitute a spinor-vector tetrad, satisfying the relations:

$$g_{\mu\nu} = \sigma_{\mu AB'} \sigma_\nu{}^{AB'}$$
$$\varepsilon_{AC} \varepsilon_{B'D'} = \sigma_{\mu AB'} \sigma^\mu{}_{CD'} , \tag{4.51}$$

where ε_{AC} and $\varepsilon_{B'D'}$ are the fundamental antisymmetric spinors in the spinor space and its complex space, respectively.

Starting with the tensor F, it is possible to construct a sequence of conserved vectors, parallel to the Maxwell electric current in electromagnetism and the Yang-Mills conserved current. The present theory is richer in conserved currents than the electromagnetic theory. Firstly, while in electromagnetism there exists only one current—the electric current, the non-Abelian nature of the gauge group in our case implies that the dual tensor $*F$ has a nonvanishing divergence, and consequently there exists a "magnetic" conserved current too. Secondly, the decomposition of the Riemann curvature tensor into its irreducible components induces a similar decomposition of the currents, with each component being conserved separately[25]. These currents are the subject of our analysis in Chapter 5.

Here we will focus our attention on a particular decomposition of the gauge field F, which owes its existence to the vectorial character of the gauge potential (compare that with the nontensorial character of the Christoffel symbols). We refer to the representation of F as a sum of a rotor part and a commutator part:

$$F_{\mu\nu} = L_{\mu\nu} + K_{\mu\nu} . \tag{4.52}$$

with

$$L_{\mu\nu} = B_{\mu,\nu} - B_{\nu,\mu} ; \qquad K_{\mu\nu} = [B_\mu, B_\nu] , \tag{4.53}$$

(spinor indices are suppressed). It is easily verified that L and K are tensors.

Consider now the tensor density $*L$ dual to the tensor L:

$$*L^{\alpha\beta} = \frac{1}{2} \varepsilon^{\alpha\beta\mu\nu} L_{\mu\nu} , \tag{4.54}$$

where ε is the Levi-Civita antisymmetric tensor density. It follows immediately from the definition that $*L$ has vanishing ordinary divergence:

$$*L^{\alpha\beta}{}_{,\beta} = 0 , \qquad (4.55)$$

i.e., it represents a conserved quantity. Notice that like the conserved complexes discussed in the previous sections, $*L$ too is derived from a superpotential:

$$*L^{\alpha\beta} = \Phi^{\alpha\beta\mu}{}_{,\mu} , \qquad (4.56)$$

with

$$\Phi^{\alpha\beta\mu} = \varepsilon^{\alpha\beta\gamma\mu} B_\gamma . \qquad (4.57)$$

We have, therefore, within the framework of the SL(2,C) gauge theory formulation of general relativity a conserved tensor density, constructed out of the metric generating matrices $\sigma_\mu{}^{AB'}$ (and their derivatives). Unlike all the other conserved complexes introduced in this chapter, the present tensor density has the advantage of being not merely an affine tensor density, but rather a tensor density with respect to the full group of coordinate transformations. Being antisymmetric, however, it suffers from the drawback associated with angular momentum.

4.7 Conclusions

In this chapter we have reviewed various attempts to solve the problem of nonconservation of the energy-momentum tensor in general relativity. The merits of the constructs due to Einstein, Landau and Lifshitz, Moller and Nissani have been discussed. The common feature of these and other approaches is a judicious definition of a complex, identified as the gravitational energy-momentum, which upon addition to the matter energy-momentum tensor, generates a complex with vanishing ordinary divergence in all coordinate systems. This complex—the total energy-momentum—is derivable from an antisymmetric superpotential.

It is generally agreed that none of these proposals can be considered as a completely satisfactory solution to the problem of energy conservation, mainly due to the nontensorial nature of the emerging expressions.

5
SL(2,C) CONSERVATION LAWS OF GRAVITATION

5.1 *Symmetry and Conservation Laws*

In the previous chapters we discussed the energy-momentum conservation problem and the variational principle in the gravitational theory. This chapter is concerned with additional conservation laws which are peculiar to curved space-time, and their relation to the SL(2,C) symmetry of the gravitational field. Originally two conservation laws were discovered within the framework of the SL(2,C) gauge theory of gravitation. The first, that of Carmeli, bears much similarity to the Yang-Mills conserved current which is associated with the isospin conservation law[13,16]. The second, that of Malin, is defined and conserved in an affine space, without the need of a metric[38].

In the next section we give a brief review of Carmeli's and Malin's conservation laws, and show, following Nissani[26], that they are members of a set of six conservation laws which are peculiar to curved space-time. In Sec. 5.3 the relationship of these conservation laws to the SL(2,C) symmetry of the gravitational field is studied. To this end we apply Noether's theorem to

Carmeli's quadratic Lagrangian[4,13,23] of the SL(2,C) gauge theory of gravitation. In Sec. 5.4 the existence of a gravitational conserved current due to Nissani[27] is proved by applying Noether's theorem with respect to the SL(2,C) symmetry to the Carmeli and Kaye[20] version of Hilbert's Lagrangian. In Sec. 5.5 we calculate this current for the Tolman metric[40]. The summary in Sec. 5.6 concludes the chapter.

5.2 *The Set of Six Currents*

In this section we exhibit a set of six conserved vector densities derived from the SL(2,C) gauge field theory of gravitation. Two of these currents were found by Carmeli and Malin, and we first focus our attention to them.

To demonstrate the existence and conservation of Carmeli's current[13,16], substitute the decomposition (2.75) of the gauge field tensor into the spinorial version of the Bianchi identities (2.56):

$$\varepsilon^{\alpha\beta\mu\nu}\left[F^W_{\mu\nu,\beta} - [B_\beta, F^W_{\mu\nu}]\right] = -\varepsilon^{\alpha\beta\mu\nu}\left[F^R_{\mu\nu,\beta} - [B_\beta, F^R_{\mu\nu}]\right]. \quad (5.1)$$

(Recall that $F^W{}_{\alpha\beta}$ and $F^R{}_{\alpha\beta}$ are 2×2 spinor matrices, whereas B_α is a 2×2 matrix undergoing nonhomogeneous SL(2,C) transformations.) The right-hand side of Eq. (5.1) is a vector density, related through the Einstein gravitational field equations to the energy-momentum tensor. As such it can be considered as the source current of the gravitational field, and is denoted by

$$\bar{J}^\alpha = -\frac{1}{2\kappa}\varepsilon^{\alpha\beta\mu\nu}\left[F^R_{\mu\nu,\beta} - [B_\beta, F^R_{\mu\nu}]\right]. \quad (5.2)$$

This vector density is not conserved. By adding to it the commutator appearing on the left-hand side of Eq. (5.1), however, we obtain Carmeli's conserved current:

$$J^\alpha_C = \kappa\bar{J}^\alpha + \frac{1}{2}\varepsilon^{\alpha\beta\mu\nu}[B_\beta, F^W_{\mu\nu}]. \quad (5.3)$$

This is a vector density with respect to general coordinate transformations of space-time, and is a 2×2 matrix subject to a nonhomogeneous SL(2,C) transformation.

The decomposition of the Riemann tensor into its irreducible components is carried out with the aid of the metric, thus the Carmeli current resides in the pseudo-Riemannian manifold rather than in the affine space[38]. As has been shown by Malin, one can define a conserved current in the affine space too, also in terms of the Carmeli gauge field tensor. This current is defined by

$$J_M^\mu = \frac{1}{2}\varepsilon^{\mu\alpha\beta\gamma}[B_\gamma, F_{\alpha\beta}] \,. \tag{5.4}$$

One can easily show, using Eqs. (5.1) and (2.56), that the Carmeli current (5.3) and the Malin current (5.4) are the divergences of the super-potentials

$$U_C^{\mu\nu} = \frac{1}{2}\varepsilon^{\mu\nu\rho\sigma}F_{\rho\sigma}^W \,, \tag{5.5a}$$

and

$$U_M^{\mu\nu} = \frac{1}{2}\varepsilon^{\mu\nu\rho\sigma}F_{\rho\sigma} \,, \tag{5.5b}$$

respectively, and whence their conservation.

Let us now consider the complete decomposition of the gauge field tensor, induced by the decomposition of the Riemann tensor into its three irreducible components, viz.,

$$F_{\mu\nu} = F_{\mu\nu}^W + F_{\mu\nu}^S + F_{\mu\nu}^T \,, \tag{5.6}$$

where F^S and F^T are the spinor counterparts of the trace-free Ricci tensor and the Ricci scalar, respectively. Since F^W, F^S and F^T are skew-symmetric tensors, one can define six super-potentials,

$$U_1^{\mu\nu} = \frac{1}{2}\varepsilon^{\mu\nu\rho\sigma}F_{\rho\sigma}^W \,,$$

$$U_2^{\mu\nu} = \frac{1}{2}\varepsilon^{\mu\nu\rho\sigma}F_{\rho\sigma}^S \,,$$

$$U_3^{\mu\nu} = \frac{1}{2}\varepsilon^{\mu\nu\rho\sigma}F_{\rho\sigma}^T \,, \tag{5.7}$$

$$U_4^{\mu\nu} = \sqrt{-g}F_W^{\mu\nu},$$

$$U_5^{\mu\nu} = \sqrt{-g}F_S^{\mu\nu},$$

$$U_6^{\mu\nu} = \sqrt{-g}F_T^{\mu\nu},$$

which generate the six conserved currents

$$J_i^\mu = U_i^{\mu\nu}{}_{,\nu} \qquad (i = 1, 2, \ldots, 6).$$

(All of them are vector densities, undergoing nonhomogeneous SL(2,C) transformations.)

The first current is Carmeli's conserved current,

$$J_C^\mu = J_1^\mu = U_1^{\mu\nu}{}_{,\nu}, \tag{5.8a}$$

whereas the sum of the first three currents is equal to Malin's conserved current,

$$J_M^\mu = J_1^\mu + J_2^\mu + J_3^\mu = U_1^{\mu\nu}{}_{,\nu} + U_2^{\mu\nu}{}_{,\nu} + U_3^{\mu\nu}{}_{,\nu}. \tag{5.8b}$$

From the viewpoint of gauge field theory the first three currents describe magnetic charges, while the last three describe electric charges of the gravitational field. Note that the magnetic charges here do not vanish, as is the case in Maxwell's field, and that is due to the non-Abelian character of the gauge group in the gravitational case.

In the next section we apply Noether's theorem to Carmeli's Lagrangian in order to establish the relationship between these currents and the SL(2,C) symmetry of the gravitational field.

5.3 *The SL(2,C) Symmetry*

In the last section we proved the existence of a set of six conserved currents emerging from the decomposition of the gravitational SL(2,C) gauge field tensor, and that this set includes Carmeli's and Malin's conserved currents. The six currents are 2 × 2 matrices which undergo nonhomogeneous SL(2,C) transformations. In this section we prove, using Noether's theorem with respect to the SL(2,C) symmetry of Carmeli's Lagrangian, the existence of an SL(2,C)-invariant version of Malin's current[23].

According to Noether's theorem, if $\mathcal{L}(Y_A, Y_{A,\mu})$ is a Lagrangian with a symmetry group which does not act on the coordinates, then

$$J_g^\mu = \sum_A \frac{\partial \mathcal{L}}{\partial Y_{A,\mu}} W_{gA} \tag{5.9}$$

is a conserved current, where

$$W_{gA} = \frac{1}{\varepsilon} \delta_g Y_A \tag{5.10}$$

is the variation of the variable Y_A induced by the infinitesimal transformation of the symmetry group,

$$S_g = I + \varepsilon g, \tag{5.11}$$

with g being a 2×2 matrix belonging to the inner space of the group.

Substituting in Eq. (5.9) Carmeli's Lagrangian (3.7), we get

$$J_g^\mu = \frac{1}{4\varepsilon} \frac{\partial \mathcal{L}_c}{\partial B_{va\ ,\mu}^{\ b}} \delta_g B_{va}^{\ b}. \tag{5.12}$$

The last expression, therefore, is a divergenceless vector density, namely, it satisfies the continuity equation

$$J_{g,\mu}^\mu = 0. \tag{5.13}$$

The variation $\delta_g B_{\alpha a}^{\ b}$ of the dyadic components of the vector gauge potential induced by an infinitesimal SL(2,C) transformation $S = (1 + \varepsilon g)$ is obtained from the nonhomogeneous SL(2,C) transformation law

$$B'_\mu = S^{-1} B_\mu S - S^{-1} S,_\mu \tag{5.14}$$

of the potentials.

According to (5.14), the conserved current (5.12) assumes the following form:

$$J_g^\nu = \text{Tr} \left[\frac{1}{2} \varepsilon^{\alpha\beta\mu\nu} [B_\mu, F_{\alpha\beta}] g + \frac{1}{2} \varepsilon^{\alpha\beta\mu\nu} F_{\alpha\beta} g,_\mu \right]. \tag{5.15}$$

Using the Bianchi identities (2.56), we get

$$J_g^\nu = \text{Tr}\left[\frac{1}{2}\varepsilon^{\alpha\beta\mu\nu}(F_{\alpha\beta}g)_{,\mu}\right]. \tag{5.16}$$

Equation (5.16) shows that this current is derived from an SL(2,C)-invariant superpotential related to the superpotential which generates Malin's current (5.5). Since g is an arbitrary matrix from the inner space of the group SL(2,C), it can be chosen so as to satisfy

$$g = g_i, \qquad g_{,\mu} = 0, \tag{5.17}$$

where g_i are the group generators. Hence we have the conserved currents

$$J_i^\mu = \text{Tr}(J_M^\mu g_i), \tag{5.18}$$

where J_M^μ is Malin's conserved current.

With the aid of the decomposition (5.6) of the gauge field tensor, Eq. (5.16) yields the three SL(2,C)-invariant currents

$$J_{1g}^\mu = \text{Tr}\left[\frac{1}{2}\varepsilon^{\alpha\beta\mu\nu}(F_{\alpha\beta}^W g)_{,\nu}\right], \tag{5.19}$$

$$J_{2g}^\mu = \text{Tr}\left[\frac{1}{2}\varepsilon^{\alpha\beta\mu\nu}(F_{\alpha\beta}^S g)_{,\nu}\right], \tag{5.20}$$

$$J_{3g}^\mu = \text{Tr}\left[\frac{1}{2}\varepsilon^{\alpha\beta\mu\nu}(F_{\alpha\beta}^T g)_{,\nu}\right]. \tag{5.21}$$

Choosing again the matrix g so as to satisfy (5.17), these currents can be related to Carmeli's current and to the currents J_2 and J_3 discussed in the previous section.

5.4 *The Gravitational Current*

In the last section an application of Noether's theorem to Carmeli's Lagrangian with the SL(2,C) symmetry was shown to produce an SL(2,C)-invariant expression for Malin's conserved current. In this section we prove, following Nissani[27], the existence of another current associated with the same symmetry and obtained from Hilbert's Lagrangian in its spinorial version.

It is based on the work of Carmeli and Kaye[20], who have expressed Hilbert's Lagrangian in terms of spinorial variables, from which they derived Einstein's gravitational field equations in spinorial form, as well as the definition of the gauge potentials.

The Carmeli-Kaye version of the Hilbert Lagrangian is given by (see Sec. 2.6)

$$\mathcal{L}_{CK} = \mathcal{L}_0(\sigma^\mu_{ab'}, B_{\mu a}{}^b, B_{\mu,\nu a}{}^b) - 2\kappa \mathcal{L}_M . \qquad (5.22)$$

(Lower-case Latin letters denote dyad-component indices of spinors with respect to a local spinor basis, and primed indices correspond to the complex conjugate spinor space.) The σ^μ are the metric tetrad tied up with the metric tensor and the fundamental skew-symmetric spinor through the relations:

$$g_{\mu\nu} = \sigma_{\mu ab'} \sigma_\nu{}^{ab'} ,$$
$$\varepsilon_{ac}\varepsilon_{b'd'} = \sigma_{\mu ab'} \sigma^\mu{}_{cd'} . \qquad (5.23)$$

The metric tetrad satisfies

$$\sigma^\mu_{AB';\nu} = (\sigma^\mu_{ab'} X^a_A \bar{X}^{b'}_{B'})_{;\nu} = 0 , \qquad (5.24)$$

where $X_A{}^b$ is the dyad basis of the spinor space. The B_μ are the vector potentials of the SL(2,C) gauge theory of gravitation defined by

$$X^A_{a;\mu} = B_{\mu a}{}^b X^A_b . \qquad (5.25)$$

The gravitational part $\mathcal{L}_0$ of the Lagrangian (5.22) is given explicitly by

$$\mathcal{L}_0 = -2\sqrt{-g}\,\sigma^{\mu\nu}{}_a{}^b F_{\mu\nu b}{}^a , \qquad (5.26)$$

where $\sigma^{\alpha\beta}$ is defined by

$$\sigma^{\alpha\beta b}_a = \sigma^{[\alpha}_{ac'} \sigma^{\beta]bc'} , \qquad (5.27)$$

(square brackets denote anti-symmetrization), and $F_{\alpha\beta}$ is the gauge field tensor given by Eq. (2.40) and discussed in Chapter 2. By definition $\sigma^{\alpha\beta}$ is a skew-symmetric tensor, whose component are 2 × 2 traceless matrices belonging to the inner space of the gauge group.

By varying the dyad components of the tetrad σ^α the spinor equivalent form of the Einstein gravitational field equations are obtained, whereas the Lagrange equations corresponding to the variation of the dyad components of the vector potential B_α define the potentials as functions of the tetrad and its derivatives. This result is analogous to the Palatini variational method when applied to Hilbert's Lagrangian, whereby the affine connections are determined by the Lagrange equations as functions of the metric tensor and its derivatives.

Since the only derivatives appearing in the gravitational part $\mathcal{L}_0$ of the Lagrangian (5.22) are in $B_{\alpha,\beta}$, and assuming that the matter Lagrangian part $\mathcal{L}_M$ does not depend on derivatives of spinorial variables, Noether's theorem [Eq. (5.9)] implies that

$$J^\mu_{\sigma i} = \frac{1}{4\varepsilon} \frac{\partial \mathcal{L}_0}{\partial B_{\nu a\ ,\mu}^{\ b}} \delta_{g_i} B_{\nu a}^{\ b} \tag{5.28}$$

is a vector density with a vanishing ordinary divergence. (The fact that it is a vector density follows from its explicit expression given below.) We have, therefore, established that this current satisfies the continuity equation

$$J^\mu_{\sigma i,\mu} = 0 . \tag{5.29}$$

The variation $\delta_{g_i} B_{\mu a}^{\ b}$ of the dyad components of the vector gauge potential induced by an infinitesimal SL(2,C) transformation $(1 + \varepsilon g_i)$ is found from the nonhomogeneous transformation law (5.14) of the potentials. Equation (5.28) then gives the following explicit form of the conserved current:

$$J^\mu_{\sigma i} = \sqrt{-g}\,\sigma^{\mu\nu\ b}_{\ \ a}([B_\nu, g_i] - g_{i,\nu})_b^{\ a} . \tag{5.30}$$

Like the other conserved vector currents enumerated in this chapter, and the conserved pseudo-tensors and tensor densities considered in Chapter 4, this current too can be derived from a superpotential. To this end we employ the identity

$$\text{Tr}\,(A[B,C]) \equiv \text{Tr}\,([A,B]C) \tag{5.31}$$

to get

$$J^\mu_{\sigma i} = \sqrt{-g}([\sigma^{\mu\nu}, B_\nu]_a^{\ b} g_{ib}^{\ a} - \sigma^{\mu\nu\ b}_{\ \ a} g_{ib}^{\ a}_{,\nu}) . \tag{5.32}$$

Now, from the definition of the vector potential (5.25) along with Eq. (5.24), it follows that

$$\sigma^{\mu\nu\,b}{}_{a\,;\rho} = [B_\rho, \sigma^{\mu\nu}]_a{}^b \,. \tag{5.33}$$

Substituting Eq. (5.33) in Eq. (5.32) casts the expression of the conserved vector density into the form of a divergence of an anti-symmetric superpotential,

$$J^\mu_{\sigma i} = -\sqrt{-g}(\sigma^{\mu\nu\,b}{}_a g_{ib}{}^a)_{;\nu} = -(\sqrt{-g}\sigma_i^{\mu\nu})_{,\nu} \,, \tag{5.34}$$

with

$$\sigma_i^{\mu\nu} = \sigma^{\mu\nu\,b}{}_a g_{ib}{}^a \,. \tag{5.35}$$

In the next section we give the explicit expression of this current in the case of the Tolman metric.

5.5 Example: The Tolman Metric

As an example for the gravitational current, the existence of which and its conservation was established in the last section, we now calculate its value in Tolman's spherically-symmetric dust-filled space-time. The energy-momentum tensor in this case is given by

$$T^{\mu\nu} = \rho u^\mu u^\nu \,, \tag{5.36}$$

where ρ is a scalar function—the rest mass density, and $u^\mu = dx^\mu/ds$ is the velocity 4-vector.

The Tolman solution of the Einstein gravitational field equations in comoving spherical coordinates $x^\mu = (t, r, \theta, \phi)$ is given by the line element[22]:

$$ds^2 = dt^2 - \frac{R'^2}{1+f(r)}dr^2 - R^2 d\theta^2 - R^2 \sin^2\theta \, d\phi^2 \,. \tag{5.37}$$

Here R is a function of t and r, satisfying

$$R' = \frac{\partial R}{\partial r} > 0 \,, \tag{5.38}$$

and

$$\dot{R}^2 = \left[\frac{\partial R}{\partial t}\right]^2 = f(r) + \frac{F(r)}{R}, \tag{5.39}$$

where $F(r)$ is defined by

$$\frac{\partial F(r)}{\partial r} = \kappa \rho R^2 R'. \tag{5.40}$$

From Eq. (5.37) it follows that R is the "radial distance" of the dust particle from the center of matter, and hence $\partial R/\partial t$ expresses the radial velocity of the particle. For $f(r) = 0$ one finds

$$\dot{R}^2 = \frac{F(r)}{R}, \tag{5.41}$$

describing a "parabolic" motion of the particles whose velocities tend to zero at infinity.

In order to find the expression of the gravitational conserved current, described in the last section, for the Tolman metric, we take the tetrad $\sigma^{(\mu)}$ in the local Minkowskian frame,

$$\sigma_{ab'}^{(0)} = \frac{I}{\sqrt{2}},$$

$$\sigma_{ab'}^{(j)} = \frac{\tau_j}{\sqrt{2}}, \tag{5.42}$$

where I is the 2×2 unit matrix, and τ_j are the Pauli matrices

$$\tau_1 = \begin{bmatrix} 0 & 1 \\ 1 & 0 \end{bmatrix},$$

$$\tau_2 = \begin{bmatrix} 0 & i \\ -i & 0 \end{bmatrix}, \tag{5.43}$$

$$\tau_3 = \begin{bmatrix} 1 & 0 \\ 0 & -1 \end{bmatrix}.$$

SL(2,C) Conservation Laws of Gravitation

As a basis for the inner space of the group SL(2,C) we choose

$$g_{ja}{}^b = \frac{T_j}{\sqrt{2}}. \tag{5.44}$$

Then the skew-symmetric tensor $\sigma^{(\alpha)(\beta)}$ in the local Minkowskian frame assumes the form:

$$\sigma_1^{(\mu)(\nu)} = \frac{1}{\sqrt{2}} \begin{bmatrix} 0 & -1 & 0 & 0 \\ 1 & 0 & 0 & 0 \\ 0 & 0 & 0 & i \\ 0 & 0 & -i & 0 \end{bmatrix},$$

$$\sigma_2^{(\mu)(\nu)} = \frac{1}{\sqrt{2}} \begin{bmatrix} 0 & 0 & -1 & 0 \\ 0 & 0 & 0 & -i \\ 1 & 0 & 0 & 0 \\ 0 & i & 0 & 0 \end{bmatrix}, \tag{5.45}$$

$$\sigma_3^{(\mu)(\nu)} = \frac{1}{\sqrt{2}} \begin{bmatrix} 0 & 0 & 0 & -1 \\ 0 & 0 & i & 0 \\ 0 & -i & 0 & 0 \\ 1 & 0 & 0 & 0 \end{bmatrix}.$$

The matrix

$$M_{(\nu)}{}^\mu = \begin{bmatrix} 1 & 0 & 0 & 0 \\ 0 & \dfrac{\sqrt{1+f}}{R'} & 0 & 0 \\ 0 & 0 & R^{-1} & 0 \\ 0 & 0 & 0 & (R\sin\theta)^{-1} \end{bmatrix}, \tag{5.46}$$

which satisfies

$$M_{(\mu)}{}^\rho M_{(\nu)}{}^\sigma \eta^{(\mu)(\nu)} = g^{\rho\sigma}, \tag{5.47}$$

will serve as the transformation matrix from the local Minkowskian coordinates to the comoving coordinates. With this choice the skew-symmetric tensor $\sigma_1{}^{\alpha\beta}$ in the comoving coordinates is cast into the form:

$$\sigma_1^{\mu\nu} = \frac{1}{\sqrt{2}} \begin{bmatrix} 0 & -\frac{\sqrt{(1+f)}}{R'} & 0 & 0 \\ \frac{\sqrt{(1+f)}}{R'} & 0 & 0 & 0 \\ 0 & 0 & 0 & \frac{i}{R^2 \sin\theta} \\ 0 & 0 & \frac{-i}{R^2 \sin\theta} & 0 \end{bmatrix} \qquad (5.48)$$

Using the relations listed above one finds the conserved vector density to be

$$J^\mu_{\sigma 1} = -(\sqrt{-g}\sigma_1^{\mu\nu})_{,\nu} = (\sqrt{2}RR' \sin\theta, -\sqrt{2}R\dot{R} \sin\theta, 0, 0) . \qquad (5.49)$$

The conserved invariant quantity, i.e., the rest "charge" density measured in the local Minkowskian frame, is then given by

$$q = (\frac{1}{-g} J^\mu_{\sigma 1} J_{\sigma 1\mu})^{\frac{1}{2}} = \frac{\sqrt{2}}{R}(1 + f - \dot{R}^2)^{\frac{1}{2}} . \qquad (5.50)$$

The coordinate velocity of this "charge" density in comoving coordinates is given by

$$\frac{dr}{dt} = \frac{J^1_{\sigma 1}}{J^0_{\sigma 1}} = -\frac{\dot{R}}{R'} . \qquad (5.51)$$

Thus the velocity of the charge density with respect to the dust particles is given by

$$v = \frac{dl}{dt} = \sqrt{-g_{11}} \frac{dr}{dt} = -\frac{\dot{R}}{\sqrt{(1+f)}} . \qquad (5.52)$$

From (5.52) and (5.50) it then follows that the charge density is equal to

$$q = \sqrt{2(1+f)} \frac{\sqrt{(1-v^2)}}{R} . \qquad (5.53)$$

5.6 Summary

In this chapter we have discussed the conservation laws which are associated with the SL(2,C) symmetry of gravitation. This symmetry provides conservation laws which are peculiar to curved space-time. All the conserved currents are vector densities, and are ordinary divergences of skew-symmetric superpotentials.

In this way we established the existence of six conserved vector currents which are formally associated with the "electric" and "magnetic" charges of the gravitational SL(2,C) gauge field. It is remarkable that gravitation exhibits properties which are characteristic to ordinary SU(2) gauge field theory, where one also encounters "electric" and "magnetic" charges. Unlike the electromagnetic theory, the "magnetic" charges do not vanish in the gravitational case. This is due to the non-Abelian nature of the gravitational gauge group, in contrast to the Abelian gauge group of the electromagnetic field. The multiplicity of these currents is a consequence of the decomposition of the Riemann tensor into its irreducible components, which is reflected by a parallel decomposition of the gauge field, the latter, in turn, induces a decomposition of the conserved currents.

Two of these six currents have been discovered previously by Carmeli and by Malin. The Carmeli conserved current is similar to the Yang-Mills current associated with isospin conservation. The Malin current is unique in its being defined and conserved in the affine space, with no need to resort to the metric.

The relationship between these conservation laws and the SL(2,C) symmetry is established with the aid of Noether's theorem as applied to Carmeli's Lagrangian. In this way an SL(2,C)-invariant form has been granted to the Malin current. The decomposition of this current, then, leads to an SL(2,C)-invariant form for the rest of the "magnetic" currents, including Carmeli's.

The application of Noether's theorem to the Carmeli-Kaye version of Hilbert's Lagrangian produces the gravitational current due to Nissani. This current is derived from an anti-symmetric superpotential equal to that of the "electric" currents, but in which the gauge field tensor is replaced by an anti-symmetric tensor constructed out of the tetrad σ^α. As an example, the explicit expression of this current, as well as the velocity associated with it, have been calculated for the case of the Tolman metric.

6
NON-ROTATING FRAMES

6.1 *Preliminaries*

Several solutions to the problem of energy conservation in general relativity, which have been proposed in the past seventy years, were reviewed in Chapter 4. All of them culminate in various forms of pseudo-tensors which are interpreted as expressions of the gravitational energy-momentum. It is generally agreed that none of them can be considered as a satisfactory solution to the underlying problem, and some of the difficulties associated with them were enumerated.

The search for a solution of the general relativistic conservation law problem hitherto has been governed by one fundamental assumption, which contrasts the conditions of ordinary versus covariant divergencelessness of the energy-momentum tensor. While the former is a manifestation of a global conservation law, the latter is perceived as expressing a conservation law valid locally only. Consequently, the thrust of the search has been to find an expression for the total energy (material plus gravitational) which should have a vanishing ordinary divergence in all coordinate systems. Thus, the expression of the total energy is required to admit a general covariant conservation law.

This chapter presents Nissani and Leibowitz's work[30,31] on preferred coordinate systems in curved space-time, in which the energy-momentum tensor is conserved globally. The study of these systems, in addition to shedding light on the issue of conservation, is of interest in its own. Investigating the features of the preferred systems is evidently of great importance in understanding their role in physical reality.

In the next section the differential equation defining the preferred frames, in which the law of energy conservation is valid globally, is derived. The interpretation of the differential equation is consistent with the Mach principle, namely, that the preferred frames are determined by the distribution of energy-momentum in spacetime. These frames are called "nonrotating", in the generalized four-dimensional sense, since given one frame of the preferred family, then in general any frame rotating with respect to it is not a member of the preferred class. Furthermore, it will be shown that the nonrotating systems are intimately related to the geodesic systems, the latter being the local geodesic inertial frames. In fact, it will be proved that the class of geodesic frames with respect to an arbitrary geodesic world line, contains a subclass of geodesic nonrotating frames. Moreover, the internal group of this subclass will be shown to be the Lorentz group together with a set of coordinate transformations which are Lorentzian up to second order in the distance from the geodesic world line. This lends support to perceiving the geodesic nonrotating systems as the inertial frames in space-time.

A concrete example, whereby the nonrotating systems can be exhibited explicitly, is given in the discussion of space-time filled with a dust of noninteracting particles. The last section of this chapter is devoted to concluding remarks.

6.2 The Nonrotating Coordinate Systems

It will now be proved that there exist preferred coordinate systems—the nonrotating coordinates—in which the energy-momentum tensor is conserved globally. To accomplish this goal, let us examine the covariant divergence-free property of the energy-momentum tensor in any given coordinate system x:

$$\sqrt{-g}T^{\mu\nu}{}_{;\nu} = (\sqrt{-g}T^{\mu\nu})_{,\nu} + \sqrt{-g}\Gamma^{\mu}_{\nu\rho}T^{\nu\rho} = 0 . \qquad (6.1)$$

This equation differs from a continuity equation by the appearance of the Γ-term. At the origin of a geodesic system the connections Γ vanish, and thus Eq. (6.1) constitutes a local continuity equation. However, it is impossible to

Non-Rotating Frames

find a coordinate system in which the connections vanish globally—the latter being 40 constraints on the four functions defining the coordinate transformation, and obviously cannot be satisfied. To reduce Eq. (6.1) to a global continuity equation, however, it is sufficient to require merely four constraints, namely,

$$\Gamma'^{\mu}_{\sigma\rho} T'^{\sigma\rho} = 0 , \qquad (6.2a)$$

where primes denote the values of the various objects in a coordinate system x', obtained from x through the coordinate transformation $x'^{\mu} = x'^{\mu}(x)$. In virtue of the Einstein field equations, (6.2a) can be written equivalently as

$$\Gamma'^{\mu}_{\sigma\rho} G'^{\sigma\rho} = 0 , \qquad (6.2b)$$

where G is the Einstein tensor.

Definition: A coordinate system x' satisfying Eq. (6.2) is called a *nonrotating system*.

Let us now translate the constraints (6.2) into a condition on the coordinate transformation $x'(x)$. From the inhomogeneous transformation rule for the Christoffel symbols,

$$\Gamma^{\rho}_{\mu\nu} = \Gamma'^{\gamma}_{\alpha\beta} \frac{\partial x'^{\alpha}}{\partial x^{\mu}} \frac{\partial x'^{\beta}}{\partial x^{\nu}} \frac{\partial x^{\rho}}{\partial x'^{\gamma}} + \frac{\partial x^{\rho}}{\partial x'^{\sigma}} \frac{\partial^2 x'^{\sigma}}{\partial x^{\mu} \partial x^{\nu}} , \qquad (6.3)$$

one obtains

$$\Gamma'^{\gamma}_{\alpha\beta} = \left[\Gamma^{\rho}_{\mu\nu} - \frac{\partial x^{\rho}}{\partial x'^{\sigma}} \frac{\partial^2 x'^{\sigma}}{\partial x^{\mu} \partial x^{\nu}} \right] \frac{\partial x^{\mu}}{\partial x'^{\alpha}} \frac{\partial x^{\nu}}{\partial x'^{\beta}} \frac{\partial x'^{\gamma}}{\partial x^{\rho}} . \qquad (6.4)$$

Substitute Eq. (6.4) in the constraints (6.2) and using the tensorial transformation of T,

$$T'^{\alpha\beta} = T^{\mu\nu} \frac{\partial x'^{\alpha}}{\partial x^{\mu}} \frac{\partial x'^{\beta}}{\partial x^{\nu}} , \qquad (6.5)$$

this yields

$$\Gamma^\rho_{\mu\nu} T^{\mu\nu} \frac{\partial x'^\gamma}{\partial x^\rho} - \frac{\partial^2 x'^\gamma}{\partial x^\mu \partial x^\nu} T^{\mu\nu} = 0 . \tag{6.6}$$

Equation (6.6) constitutes a set of four unlinked linear partial differential equations for the four transformation functions $x'(x)$. Equivalently, in virtue of the Einstein field equations, this can be written in the form:

$$\Gamma^\rho_{\mu\nu} G^{\mu\nu} \frac{\partial x'^\gamma}{\partial x^\rho} - \frac{\partial^2 x'^\gamma}{\partial x^\mu \partial x^\nu} G^{\mu\nu} = 0 . \tag{6.7}$$

Notice that (6.7) is none other than the equation defining the harmonic coordinates, with the metric tensor replaced by the Einstein tensor.

The coordinate frames x' defined by Eq. (6.7) are distinguished by the property that in these frames the energy-momentum tensor density $\sqrt{-g'}T'$ satisfies globally an ordinary divergencelessness condition:

$$(\sqrt{-g'}T'^{\mu\nu})_{,\nu} = 0 . \tag{6.8}$$

It follows from this global continuity equation, in the usual manner, that for an isolated physical system, confined to a spatial volume V' (i.e., $T' = 0$ outside V'), the integrals

$$P'^\mu = \int_{V'} \sqrt{-g'}\, T'^{\mu 0} d^3 x' \tag{6.9}$$

define four physical quantities, which are conserved in time, and behave as a vector under the action of the affine group.

In the next section it will be shown that the coordinate systems x' can be specialized further, so that in addition to being nonrotating they are also geodesic with respect to the world line of the observer. In these frames, the quantities P'^μ defined by (6.9), are precisely the components of the energy-momentum vector of the system as measured in special relativity. Notice that the integration volume V' is not restricted in its size—in these nonrotating geodesic frames, unlike general geodesic frames, the integrals (6.9) are strictly conserved even for a finite volume.

At this point, some elaboration of the name nonrotating, given to the preferred coordinate systems, is in order. Let x be a geodesic nonrotating frame. In the next section it will be demonstrated that the internal group of the nonrotating

Non-Rotating Frames

frames contains the group of affine coordinate transformations. Consequently, if x' is a coordinate system obtained from x by the Lorentz transformation

$$x'^0 = \gamma(x^0 - \beta x^1),$$

$$x'^1 = \gamma(x^1 - \beta x^0), \qquad (6.10)$$

$$x'^2 = x^2,$$

$$x'^3 = x^3,$$

with $\gamma = 1/\sqrt{(1 - \beta^2)}$ and $\beta < 1$ being a constant, then x' too is a geodesic nonrotating system. The components P'^μ of the energy-momentum affine vector in the frame x', viz.,

$$P'^0 = \gamma(P^0 - \beta P^1),$$

$$P'^1 = \gamma(P^1 - \beta P^0), \qquad (6.11)$$

$$P'^2 = P^2,$$

$$P'^3 = P^3,$$

are conserved. On the other hand, if x' is a coordinate system obtained from x by a "time-dependent Lorentz transformation", i.e., the transformation (6.10) with β being a function of time, then clearly P'^μ will no longer be constant in time. In view of (6.1), the last observation entails that in the coordinate system x' Eq. (6.2) does not hold. Thus we have illustrated the fact that, in general, a coordinate system in a state of rotation with respect to a "nonrotating" system does not belong to the class of "nonrotating" frames, justifying the name given to this preferred class of frames.

The group of transformations connecting nonrotating frames will be derived in the next section, and it will be shown that the Lorentz group, which leaves the class of Minkowskian-nonrotating-geodesic frames invariant, is contained in it.

6.3 The Internal Group

In the last section, Eq. (6.7) was shown to be the differential equation defining the transformation from an arbitrary coordinate system x to a nonrotating system

x'. We now write down the differential equation defining the internal group of the class of nonrotating frames.

Let x be a nonrotating coordinate system, namely, a system satisfying

$$\Gamma^\mu_{\sigma\rho} T^{\sigma\rho} = 0, \qquad \Gamma^\mu_{\sigma\rho} G^{\sigma\rho} = 0 \,,$$

[cf. Eq. (6.2)]. A necessary and sufficient condition for a coordinate system x' to be nonrotating, according to the discussion of the last section, is given by Eq. (6.6) or (6.7). Combining them with the last equations, we find the necessary and sufficient condition for a transformation function to belong to the internal group of the nonrotating frames

$$T^{\mu\nu} \frac{\partial^2 x'^\gamma}{\partial x^\mu \partial x^\nu} = 0 \,, \qquad (6.12)$$

or equivalently

$$G^{\mu\nu} \frac{\partial^2 x'^\gamma}{\partial x^\mu \partial x^\nu} = 0 \,. \qquad (6.13)$$

Evidently, this internal group contains the affine group as a proper subgroup. This subgroup transforms the nonrotating frames which are geodesic with respect to a given point or a world line among themselves. (The existence of such frames will be proved in the next section.) The nonlinear transformations belonging to the internal group serve as the vehicle for transforming from a nonrotating frame, geodesic with respect to an observer, to a nonrotating frame, geodesic with respect to another observer. It follows that the internal group of the class of frames which are nonrotating-Minkowskian-geodesic with respect to an observer consists of the Lorentz group and a set of transformations which are Lorentzian up to second order. This class, therefore, can be viewed as the inertial frames of curved space-time. Unlike the situation in special relativity, in curved space-time an inertial frame relative to an observer, is not necessarily inertial (though it is nonrotating) relative to another observer.

It is worth noticing that Eq. (6.13) is precisely d'Alembert's equation, with the Einstein tensor replacing the metric tensor (cf. the relation between Eq. (6.7) and the harmonic coordinates, alluded to in the previous section).

References were made in the last sections to coordinate systems which are both nonrotating and geodesic. Their existence will be proved in the next section.

6.4 Inertial Coordinates

Nonrotating frames were defined in the previous sections as those coordinate systems in which the energy-momentum tensor is conserved globally. It has been proved that any space-time manifold admits nonrotating frames. We will now show that given an arbitrary geodesic world-line A, and a coordinate system x geodesic with respect to A, then it is always possible to find transformation functions from x to a coordinate system x' which is nonrotating and geodesic with respect to A, and which furthermore coincides with x up to second order in the distance from A:

$$x'^{\mu} - x^{\mu} = a_i^{\mu}(x^i - x_A^i)^3 + \ldots, \qquad (6.14)$$

(where the subscript A denotes values along the line A, and a_i are constants). The geodesic nonrotating coordinate systems will be called *the inertial frames of curved space-time*.

To prove the assertion, let x be a coordinate system which is geodesic with respect to A, viz.,

$$\Gamma^{\gamma}_{A\alpha\beta} = 0, \qquad (6.15)$$

and $x'(x)$ the transformation functions from x to a coordinate system x' (which will be required to be nonrotating and geodesic). Consider the Taylor expansion (in the deviation from A) of the function x':

$$x'^{\gamma} = x_A^{\gamma} + \left.\frac{\partial x'^{\gamma}}{\partial x^i}\right|_A x^i + \frac{1}{2}\left.\frac{\partial^2 x'^{\gamma}}{\partial x^i \partial x^j}\right|_A x^i x^j + \ldots, \qquad (6.16)$$

where Latin indices assume the values 1,2,3. For convenience we have chosen $x_A' = x_A$ and $x_A^i = 0$. Our goal is to show that the coefficient of this series can be determined in such a way that the coordinate system x' becomes geodesic and nonrotating.

First we set

$$\left.\frac{\partial x'^{\gamma}}{\partial x^{\alpha}}\right|_A = \delta^{\gamma}_{\alpha}, \qquad (6.17)$$

$$\left.\frac{\partial^2 x'^{\gamma}}{\partial x^i \partial x^j}\right|_A = 0. \qquad (6.18)$$

This choice implies that the condition (6.7) for the nonrotating nature of x' is satisfied at least along A. It also establishes the geodesic character of x' and, furthermore, it entails that the difference between x and x' will be at most of the third order in the deviation from the line A. Now, taking the values of the successive derivatives of both sides of Eq. (6.7) along the line A, one obtains a set of algebraic equations in the third and higher order coefficients of (6.16), which must be satisfied in order that x' will be nonrotating. It is a straightforward, though tedious, matter to write down the explicit conditions, but it will be omitted here, since the same procedure is carried out in detail in the alternative derivation which follows [Eqs. (6.19)–(6.24)]. For the same reason we ignore at this point the question regarding the convergence of the series. We do stress, however, that the coefficients obtained by solving the algebraic equations are determined solely by the local values of the geometric parameters of space-time along the world line A.

Conversely, suppose that x is a nonrotating, but nongeodesic coordinate system. From the internal group defined by Eq. (6.13) of the class of nonrotating frames we can select a coordinate transformation $x'(x)$ subject to the additional constraints

$$\left.\frac{\partial x'^{\gamma}}{\partial x^{\sigma}}\right|_A \Gamma^{\sigma}_{A\alpha\beta} = \left.\frac{\partial^2 x'^{\gamma}}{\partial x^{\alpha}\partial x^{\beta}}\right|_A . \tag{6.19}$$

Then, using the transformation law of the Christoffel symbols, one finds that the latter vanish along A in the coordinates x', hence x' is a nonrotating frame, geodesic with respect to A. Notice the compatibility of the constraints (6.19) with Eq. (6.13): the contractions of both sides of Eq. (6.19) with Einstein tensor,

$$\left.G_A^{\alpha\beta}\frac{\partial^2 x'^{\gamma}}{\partial x^{\alpha}\partial x^{\beta}}\right|_A = \left.G_A^{\alpha\beta}\Gamma^{\sigma}_{A\alpha\beta}\frac{\partial x'^{\gamma}}{\partial x'^{\sigma}}\right|_A , \tag{6.20}$$

vanish—the left-hand side as a result of Eq. (6.13), and the right-hand side as a result of (6.2).

To complete the proof, it is necessary to verify the existence of such a transformation. To do that explicitly, we write the transformation functions in the following form:

$$x'^{\alpha} = x_A^{\alpha}(x^0) + \left.\frac{\partial x'^{\alpha}}{\partial x^i}\right|_A (x^0) x^i + \frac{1}{2}\left.\frac{\partial^2 x'^{\alpha}}{\partial x^j \partial x^k}\right|_A (x^0) x^j x^k + F^{\alpha}_{ijk} x^i x^j x^k . \tag{6.21}$$

Non-Rotating Frames

Here F denotes the sum of all the third and higher order terms in the Taylor expansion. For x' to be geodesic, the first- and second-order terms in Eq. (6.21) are required to satisfy Eq. (6.19), which decomposes into two sets of conditions:

$$\frac{d}{dx^0}\left[\frac{\partial x'^\alpha}{\partial x^j}\bigg|_A(x^0)\right] = \frac{\partial x'^\alpha}{\partial x^\sigma}\bigg|_A(x^0)\Gamma^\sigma_{A0j}(x^0) , \qquad (6.22a)$$

$$\frac{\partial^2 x'^\alpha}{\partial x^i \partial x^j}\bigg|_A(x^0) = \frac{\partial x'^\alpha}{\partial x^\sigma}\bigg|_A(x^0)\Gamma^\sigma_{Aij}(x^0) . \qquad (6.22b)$$

For each value of α, Eq. (6.22a) constitutes a system of three ordinary differential equations of the first order in the three function $(\partial x'^\alpha/\partial x^j)_A$. (Recall that we assume $(\partial x'^\alpha/\partial x^0)_A = \delta_0^\alpha$.) Equation (6.22b), which is a set of algebraic conditions, determines the second-order coefficients $(\partial^2 x'^\alpha/\partial x^j \partial x^k)_A$.

So far, the first and second order coefficients have been fixed, thereby establishing the geodesic nature of the frame x'. Now the functions F will be determined by the requirement that the transformation $x'(x)$ is a member of the internal group of the class of nonrotating frames, i.e., that the transformation $x'(x)$ satisfies Eq. (6.13). Taking into account Eq. (6.22), this requirement assumes the form:

$$G^{00}\left\{\frac{d}{dx^0}\left[\frac{\partial x'^\gamma}{\partial x^\sigma}\bigg|_A \Gamma^\sigma_{Ai0}\right]x^i + \frac{1}{2}\frac{d^2}{dx^{02}}\left[\frac{\partial x'^\gamma}{\partial x^\sigma}\bigg|_A \Gamma^\sigma_{Aij}\right]x^ix^j\right.$$

$$\left.+ \frac{\partial^2 F^\gamma_{ijl}}{\partial x^{02}}x^ix^jx^l\right\} + G^{0i}\left\{\frac{\partial x'^\gamma}{\partial x^\sigma}\bigg|_A \Gamma^\sigma_{Ai0} + \frac{d}{dx^0}\left[\frac{\partial x'^\gamma}{\partial x^\sigma}\bigg|_A \Gamma^\sigma_{Aij}\right]x^j\right.$$

$$\left.+ \frac{\partial^2 F^\gamma_{ljm}}{\partial x^0 \partial x^i}x^lx^jx^m + 3\frac{\partial F^\gamma_{ijm}}{\partial x^0}x^jx^m\right\} + G^{ij}\left\{\frac{\partial x'^\gamma}{\partial x^\sigma}\bigg|_A \Gamma^\sigma_{Aij}\right.$$

$$\left.+ \frac{\partial^2 F^\gamma_{lmp}}{\partial x^i \partial x^j}x^lx^mx^p + 6\frac{\partial F^\gamma_{ijp}}{\partial x^i}x^lx^p + 6F_{ijp}x^p\right\} = 0 . \qquad (6.23)$$

For each value of γ the last condition is a single second-order partial differential equation for the ten functions F, so that a great amount of freedom in their choice still remains.

To verify that Eq. (6.23) admits analytic solutions, notice that in virtue of the line A being geodesic, $\Gamma_A{}^\sigma{}_{00} = 0$ (recall that $x_A{}^i = 0$), and consequently Eq. (6.23) can be cast in the form

$$G^{\alpha\beta}(x)\Gamma^\sigma_{A\alpha\beta}(x^0)\frac{\partial x'^\gamma}{\partial x^\sigma}\bigg|_A (x^0) + x^j H^\gamma_{\alpha\beta j}(x)G^{\alpha\beta}(x) = 0 , \qquad (6.24)$$

where H are expressions in the functions F and their derivatives. The second term in Eq. (6.24) vanishes for $x^j = 0$ (i.e., along the line A). It follows that the condition for the existence of analytic solutions for H is precisely the vanishing of the right-hand side of Eq. (6.20). The latter is satisfied since x is a nonrotating frame, and therefore subject to the equation

$$\Gamma^\mu_{\sigma\rho}G^{\sigma\rho} = 0 .$$

This establishes the convergence of the Taylor expansion (6.21) in a suitable neighborhood of the line A.

To sum up, the foregoing discussion shows that there exist coordinate systems which are both nonrotating and geodesic with respect to an arbitrary geodesic line A. Moreover, these frames can be obtained by applying an appropriate coordinate transformation to any geodesic frame (with respect to A), so that:
(a) The difference between the resulting geodesic nonrotating system and the original geodesic system is of third order in the deviation from the line A.
(b) The geodesic nonrotating system obtained thereby for a given time x^0 depends on geometric parameters of the underlying space-time manifold, but only on the values of the latter along the line A at the same time x^0.

The nonrotating frames, analyzed so far in the case of general curved space-time, will be illustrated in the next section in a special case of a universe permeated by a cloud of dust particles.

6.5 Example

In the previous sections a general existence proof for nonrotating frames was given. In these frames an energy-momentum conservation law holds globally. Physical insight into the nature of these preferred coordinate systems was gained by studying the relation between them and the geodesic systems. In the present section a concrete example will be given, in the special case of space-time with an energy-momentum tensor corresponding to a cloud of dust particles, but other than that with an arbitrary metric.

Consider a dust cloud of noninteracting particles. It is characterized by a scalar $\rho(x)$—the proper density, and a unit 4-vector $u^\mu(x)$—the 4-velocity field. The energy-momentum tensor of this physical set-up is

$$T^{\mu\nu} = \rho u^\mu u^\nu. \tag{6.25}$$

The covariant divergence of the energy-momentum tensor vanishes. Hence

$$\rho u^\mu{}_{;\nu} u^\nu + u^\mu (\rho u^\nu)_{;\nu} = 0. \tag{6.26}$$

Contracting this equation with u_μ gives

$$(\rho u^\nu)_{;\nu} = 0. \tag{6.27}$$

Substituting this equation into Eq. (6.26) yields

$$u^\mu{}_{;\nu} u^\nu = 0, \tag{6.28}$$

which means that the trajectories of the dust particles are geodesic lines of the space-time manifold. With the aid of Eq. (6.25), the last equation can be written in the form

$$u^\mu{}_{,\nu} u^\nu + \rho^{-1} \Gamma^\mu_{\nu\gamma} T^{\nu\gamma} = 0. \tag{6.29}$$

One has therefore the corollary that a coordinate system, in the present situation, is a nonrotating frame if and only if the 4-velocity u^μ satisfies

$$u^\mu{}_{,\nu} u^\nu = 0. \tag{6.30}$$

In particular, a "comoving frame", defined by

$$u^\mu = \delta_0{}^\mu, \tag{6.31}$$

is nonrotating.

Having introduced one particular nonrotating frame x, the most general nonrotating frame x' is derived from x by applying to it the internal group defined by Eq. (6.12). In view of Eqs. (6.25) and (6.31), the differential equation of the internal group (6.12) reduces to the simple equation

$$\frac{\partial^2 x'^{\gamma}}{\partial x^{02}} = 0 \ . \tag{6.32}$$

The general solution of this equation is

$$x'^{\mu} = x^0 A^{\mu}(x^i) + B^{\mu}(x^i) \ , \tag{6.33}$$

where A^{μ} and B^{μ} are arbitrary functions of the spatial coordinates. The transformation (6.33), therefore, generates all the nonrotating frames when applied to the comoving frame x. For constants A^{μ}, and linear functions B^{μ}, Eq. (6.33) defines the affine group, which is, as remarked in Sec. 6.3, a subgroup of the internal group.

To construct a frame which will be both nonrotating and geodesic with respect to a given point or line A, apply to the comoving frame the transformation (6.33), where now the functions A^{μ} and B^{μ} are subject to the constraints (6.19), viz.,

$$A^{\mu}{}_{,i}|_A = \Gamma^{\mu}_{A0i} \ ,$$
$$x^0 A^{\mu}{}_{,ij}|_A + B^{\mu}{}_{,ij}|_A = \Gamma^{\mu}_{Aij} \ . \tag{6.34}$$

Thus, we have exhibited explicitly the class of nonrotating frames in any space-time with underlying metric consistent with the special energy-momentum tensor (6.25). It has also been demonstrated how a nonrotating coordinate system can be produced, which will be geodesic with respect to a given world line.

6.6 Conclusions and Remarks

We have shown that there exists a family of preferred coordinate systems in which the ordinary divergence of the energy-momentum tensor vanishes, i.e., the ordinary continuity equation for the energy-momentum tensor holds in a finite domain, and not merely in an infinitesimal region as is the case in general geodesic frames. These preferred frames are entitled "nonrotating", since in general any coordinate system which is in a state of rotation with respect to these frames is not a member of the preferred class.

It follows from the differential equation (6.6), which defines the nonrotating frames, that the distribution of matter throughout space-time determines whether or not a coordinate system is nonrotating. In empty regions of space-time, where the energy-momentum vanishes, the defining equation becomes an identity.

The internal group of the class of nonrotating frames is defined by the d'Alembert-like equation (6.13). This group contains the subgroup of the affine transformations, which acts invariably on the subset of nonrotating frames which are geodesic with respect to a given line. In addition, the internal group includes the nonaffine transformations connecting nonrotating frames which are geodesic with respect to distinct lines. It follows that the internal group of the class of geodesic-Minkowskian-non-rotating frames consists of the Lorentz group along with the transformations which are Lorentzian up to second order. This observation gives further support to perceiving these frames as the inertial frames of curved space-time.

REFERENCES

1. A. Einstein, *Ann. Phys.* **49** (1916) 769.
2. M. Carmeli, *Nuovo Cimento Lett.* **4** (1970) 40.
3. M. Carmeli, *J. Math. Phys.* **11** (1970) 2728.
4. M. Carmeli and S.I. Fickler, *Phys. Rev.* **D5** (1972) 290.
5. M. Carmeli, *Nuovo Cimento* **7A** (1972) 9.
6. M. Carmeli, *Nucl. Phys.* **38B** (1972) 621.
7. M. Carmeli, *Ann. Phys. (N.Y.)* **71** (1972) 603.
8. M. Carmeli, *Gen. Relativ. Grav.* **3** (1972) 317.
9. M. Carmeli, *Gen. Relativ. Grav.* **5** (1974) 287.
10. M. Carmeli, "SL(2,C) symmetry of the gravitational field," in *Group Theory in Nonlinear Problems*, ed. A.O. Barut (D. Reidel Publishing Co., Dordrecht, Holland/Boston, 1974).
11. M. Carmeli, *Phys. Rev. Lett.* **36** (1976) 59.
12. M. Carmeli and M. Kaye, *Nuovo Cimento* **34B** (1976) 225.
13. M. Carmeli, *Phys. Rev.* **D14** (1976) 1727.
14. M. Carmeli and M. Kaye, *Nuovo Cimento Lett.* **17** (1976) 275.
15. M. Carmeli, *Phys. Rev.* **D14** (1976) 2518.
16. M. Carmeli, *Nuovo Cimento Lett.* **18** (1977) 17.
17. M. Carmeli and S. Malin, *Ann. Phys. (N.Y.)* **103** (1977) 208.

18. M. Carmeli and M. Kaye, *Nuovo Cimento* **39B** (1977) 187.
19. M. Carmeli, *Group Theory and General Relativity* (McGraw-Hill, New York, 1977).
20. M. Carmeli and M. Kaye, *Ann. Phys. (N.Y.)* **113** (1978) 177.
21. M. Kaye, *Nuovo Cimento* **43B** (1978) 293.
22. M. Carmeli, *Classical Fields* (Wiley, New York, 1982).
23. M. Carmeli and N. Nissani, *Phys. Lett.* **B113** (1982) 375.
24. N. Nissani, *Nuovo Cimento* **78A** (1983) 378.
25. N. Nissani, *Nuovo Cimento Lett.* **39** (1984) 289.
26. N. Nissani, *Phys. Reports* **109** (1984) 96.
27. N. Nissani, *Int. J. Theor. Phys.* **24** (1985) 675.
28. N. Nissani, *Phys. Rev.* **D31** (1985) 1489.
29. N. Nissani, "A modified quadratic Lagrangian for general relativity theory," in *Proceedings of the Fourth Marcel Grossmann Meeting on General Relativity*, ed. R. Ruffini (Elsevier Science Publisher B.V., Amsterdam, 1986), p.809.
30. N. Nissani and E. Leibowitz, *Phys. Lett.* **A126** (1988) 447.
31. N. Nissani and E. Leibowitz, *Int. J. Theor. Phys.* **28** (1989) 235.
32. D. Hilbert, *Konigl. Gesell. d. Wiss., Göttingen, Nachr., Math. Phys. Kl.* 395 (1915).
33. C.N. Yang and R.L. Mills, *Phys. Rev.* **96** (1954) 191.
34. M. Soffel, B. Muller and W. Greiner, *Phys. Reports* **85** (1982) 51.
35. E.T. Newman and R. Penrose, *J. Math. Phys.* **3** (1962) 566.
36. L. Herrera, *Nuovo Cimento* **17A** (1973) 48.
37. L. Herrera, *Nuovo Cimento Lett.* **21** (1978) 11.
38. S. Malin, *Nuovo Cimento* **B39** (1977) 319.
39. E. Noether, *Goett. Nachr.*, 235 (1918).
40. R.C. Tolman, *Proc. Nat. Acad. Sci.* **20** (1934) 169.
41. J.L. Anderson, *Principles of Relativity Physics* (Academic Press, New York, 1967).
42. W.L. Bade and H. Jehle, *Rev. Mod. Phys.* **25** (1953) 714.
43. E. Leibowitz, *Phys. Rev.* **D15** (1977) 2139.
44. A. Palatini, *Rend. Circ. Mat. Palermo* **43** (1919) 203.
45. P. Havas, *Gen. Relativ. Grav.* **8** (1977) 631.
46. C. Lanczos, *Ann. Math.* **39** (1938) 842.
47. H.A. Buchdahl, *Proc. Edinburgh Math. Soc.* **8** (1948) 89.
48. R. Weitzenbock, *Sitzber. Akad. Wiss. Wien.* **130, 11a** (1921) 15.
49. H. Weyl, *Ann. Phys.* **59** (1919) 101.
50. A. Eddington, *The Mathematical Theory of Relativity* (Cambridge University Press, London, 1924).
51. A. Einstein, *S. B. Preuss. Akad. Wisss.*, 778 (1915).
52. A. Einstein, *S. B. Preuss. Akad. Wisss.*, 1115 (1916).
53. A. Einstein, *S. B. Preuss. Akad. Wisss.*, 1111 (1916).
54. E. Schrödinger, *Phys. Z.* **19** (1918) 4.
55. F. Klein, *Nachr. Gesell. Wiss. Gott. Math. Phys. Kl.*, 394 (1918).
56. C. Moller, *Mat. Fys. Medd. Dan. Vid. Selsk.* **1**, No. 10 (1961).
57. A. Einstein, *Jahrb. Rad. Elektr.* **4** (1907) 411.
58. A., Einstein, *Ann. Phys.* **38** (1912) 355.

References

59. F. I. Cooperstock, *Ap. J.* **291** (1985) 460.
60. A. Einstein, *Phys. Z.* **19** (1918) 115.
61. C. Moller, *Ann. Phys. (N.Y.)* **12** (1961) 118.
62. C. Moller, *Mat. Fys. Medd. Dan. Vid. Selsk.* **35**, No. 3 (1966).
63. E. Schmutzer, *Relativistische Physik* (Teubner, Leipzig, 1968), p. 874.
64. L.D. Landau and E.M. Lifshitz, *The Classical Theory of Fields* (Pergamon Press, Oxford, 1975).
65. H. Bauer, *Phys. Z.* **19** (1918) 163.
66. Ph., Von Freud, *Am. Math. J.* **40** (1939) 417.
67. C. Moller, *Ann. Phys. (N.Y.)* **4** (1958) 347.
68. C. Moller, *Mat. Fys. Medd. Dan. Vid. Selsk.* **31**, No. 14 (1959).
69. N. Rosen, *Phys. Rev.* **57** (1940) 147.
70. N. Rosen, *Ann. Phys. (N.Y.)* **22** (1963) 1.
71. C. Moller, *Nucl. Phys.* **57** (1964) 330.
72. F.H.J. Cornish, *Proc. Roy. Soc.* **A282** (1964) 358.

SUBJECT INDEX

Action integral, 30
Action principle, 31, 42
Affine connections, 3, 29, 31
   spinor, 8, 9, 10, 11, 12, 14
Affine group, 51, 53, 80, 88
Affine space-time,
   spinor, 2
Algebra,
   spinor, 4–7
Angular-momentum complex, 55
Anti-commutation relations, 25
Antisymmetric conserved tensor density, 60–62
Astrophysics, 1

Base space of a fiber bundle, 2
Basis,
   spinor, 7, 11, 12, 21
Bauer, H., 56, 58, 60
Bianchi identities, 14, 15, 16, 17, 20, 21, 64, 68
   contracted, 47

Bi-metric theory, 58
Birkhoff theorem, 32

Carmeli, M., 2, 4, 33, 63, 64, 69, 75
Carmeli conservation law, 63
Carmeli current, 2, 64, 65, 66, 68, 75
Carmeli gauge field tensor, 65
Carmeli Lagrangian, 2, 30, 33–34, 35, 42, 64, 66, 67, 68, 75
Carmeli SL(2, C) gauge theory of gravitation, 3–27, 30, 46, 60
Carmeli-Kaye Lagrangian, 2, 26, 27, 64, 69, 75
Calculus,
   spinor, 5
   tensor, 5
Christoffel symbols, 3, 30, 36, 47, 79
Commutator of covariant derivatives, 13, 14
Comoving coordinates, 71, 73, 74
Completeness relations, 25
Complex,

angular-momentum, 55
  Einstein, 49–53
  Landau-Lifshitz, 53–56
  Moller, 56–60
Conformal tensor, 18, 19, 20, 31
Conjugate spinor space, 5
Connecting mixed quantities, 9
Connections,
  affine, 3, 29, 31
  spinor affine, 8, 9, 10, 11, 12, 14
Conservation law,
  Carmeli's, 63
  Malin's, 63
Conservation laws, 1, 2
  symmetry and, 63–64
Conservation laws associated with the SL(2, C) gauge theory of gravitation, 2, 46
Conservation laws in general relativity, 1, 2, 46
Conservation laws of gravitation, SL(2, C), 63–75
Conservation of energy-momentum, 45, 46
Conservation problem in general relativity, energy, 45–62
Conserved tensor density — gauge formulation approach, 60–62
Conserved gravitational current, 2
Conserved skew-symmetric tensor density, 2
Conserved tensor density,
  Nissani's, 46
Continuity equation, 45, 48, 67, 78, 79
  global, 2
Contracted Bianchi identity, 47
Contravariant spinor, 5
Coordinate system,
  geodesic, 4, 46, 47, 78
  preferred, 2, 78
Coordinate systems, 46
  equivalence of, 3
Coordinates,
  comoving, 71, 73, 74
  harmonic, 80
  inertial, 83–86
  Minkowskian, 9
Coordinate transformations, 7

group of, 7
Coupling with matter, 20–21
Correspondence between spinors and tensors, 4
Covariance, 3
Covariant derivative, 3, 8, 24
  commutator of, 13, 14
  dyad components of, 24
Covariant divergence, 45, 46
Covariant spinor, 5, 7
Current,
  Carmeli, 2, 64, 65, 66, 68, 75
  conserved, 2
  gravitational, 2, 68–71
  Malin, 2, 64, 65, 66, 68, 75
  Nissani, 4, 64, 69, 75
  Yang-Mills, 63, 75
Curvature, 14
Curvature spinor-tensor, 14
Curvature tensor, 2, 3, 15, 16–19, 20, 29, 30, 31
Curved space-time, 5, 10

D'Alembert equation, 82
D'Alembert-like equation, 89
Decomposition of the Riemann tensor, 18, 65
Densities,
  scalar, 32, 36, 37
Density,
  conserved skew-symmetric tensor, 2
  electromagnetic Lagrangian, 19
  free-field gravitational Lagrangian, 19
  gravitational field, 21
  Lanczos scalar, 35
Derivation of SL(2, C) gauge field equations by Palatini method, 25–27
Derivative,
  covariant, 3, 8, 24
  partial, 24
Divergence,
  covariant, 45, 46
  ordinary, 45, 46
Double-valued representation, 4
Dust, 78, 87
Dyad, 7
Dyad components of:
  covariant derivative, 24

Subject Index

partial derivative, 24
spinor, 11, 12
trace-free Ricci spinor, 22
Weyl spinor, 22
Dyad indices, 11
Dynamical law, 19, 31

Einstein A., 46, 48, 49, 50, 51, 62
Einstein complex, 49–53
Einstein field equations, 1, 18, 25, 29, 30, 31, 38, 42, 43, 47
Einstein general relativity, 1, 32
Einstein gravitational constant, 37
Einstein Lagrangian, 50, 59
Einstein pseudo-tensor, 49–53
Einstein solution, 49-53
Einstein tensor, 30, 47, 54
Einstein theory of gravitation, 2, 4
Electromagnetic stress-energy tensor, 48
Energy,
  gravitational, 46, 48
  matter, 48
Energy conservation problem in general relativity, 45–62
Energy-momentum conservation, 45, 46
Energy-momentum tensor, 2, 18, 31, 36, 38, 45, 48, 51, 78, 80, 87, 88
Energy-momentum vector, 53
Equation,
  continuity, 45, 48, 67, 78, 79
  d'Alembert, 82
  d'Alembert-like, 89
  Lagrange, 20, 21, 27, 38, 49, 50, 52
  metric, 21, 25
  Poisson, 32
Equations,
  Einstein, 1, 18, 25, 29, 30, 31, 38, 42, 43, 47
  Maxwell, 19
  Newman-Penrose, 4, 25, 33, 34, 42
Equivalence of coordinate systems, 3
Equivalence principle, 4, 46
Example, 86–88
Example: Tolman metric, 71–74
Extension of Palatini's method, 25

Fiber bundle,
  base space of, 2

Fiber bundle formulation of the $SL(2, C)$ gauge theory of gravitation, 2
Field,
  gauge, 1, 4, 13–16
  tetrad, 58, 59
Field equations, 19–21
  Einstein, 1, 18, 25, 29, 30, 31, 38, 42, 43, 47
  gravitational, 4, 18
Field theory, 1
  $SU(2)$ gauge, 42
Fierro, Martin, 1, 29
Flat four-dimensional manifold, 5
Flat space-time, 9
Formalism,
  Newman-Penrose, 2
Four-index energy-momentum tensor, 36
Frame,
  spinor, 11
Frames,
  inertial, 83
  non-rotating, 77–89
Free-field equations, 19–20
Free-field Lagrangian density, 19, 26
Fundamentals of the $SL(2, C)$ gauge theory of gravitation 3–27

Gauge field, 1, 4, 13–16
  transformation law of, 13
Gauge field tensor, 65, 68
Gauge field theory,
  $SU(2)$, 42
Gauge formulation approach to conserved tensor density, 60–62
Gauge potential, 4, 11–12
  transformation law of, 12
Gauge theory, 1
  Carmeli's $SL(2, C)$, 3–27, 60
  MMG, 4
  non-Abelian, 3
  Yang-Mills, 8, 19
Gauge theory of gravitation, 1, 2
  Carmeli's quadratic Lagrangian for, 30
  fundamentals of the $SL(2, C)$, 3–27
  $SL(2, C)$, 1, 2, 3–27
General covariance principle, 3, 45, 46
General relativity, 1, 4, 32
  conservation laws in, 1, 2, 46

Einstein's 1, 32
  energy conservation problem in, 45–62
  formulation in terms of the group SL(2, C), 4
  quadratic Lagrangian formulation of, 2
Geodesic coordinate systems, 4, 46, 47, 78
Global continuity equation, 2
Gravitation, 1
  Einstein's theory of, 2, 4
  gauge theory of, 1, 2
  SL(2, C) conservation laws of, 63–75
  SL(2, C) gauge theory of, 1, 2, 3–27, 30, 46
Gravitation as an SL(2, C) gauge theory, 11–19
Gravitational constant, Einstein, 37
Gravitational current, 2, 68–71
Gravitational energy, 46, 48
Gravitational field,
  quantization program of, 42
  source of, 21
Gravitational field equations, 4, 18, 30, 31, 38
Gravitational field Lagrangian density, 21, 27
Gravitational Lagrangian, 5, 51
Gravitational tensor, 36
Group,
  affine, 51, 53, 80, 88
  general coordinate transformations, 3, 7
  internal, 18, 81–82, 88, 89
  Lorentz, 4, 8, 78, 81, 82
  manifold mapping (MMG), 3, 7, 8, 17
  proper (homogeneous) Lorentz, 4
  representations of the Lorentz, 5
  SL(2, C), 2, 4, 5, 7, 12, 42

Harmonic coordinates, 80
Havas, P., 32
Hermitian spinor, 10
Hermitian spinor vector, 9
Hernandez, Jose, 1, 29
High-order theories, 32
High-order theories Lagrangians, 32
Hilbert, D., 1, 29, 30
Hilbert's Lagrangian, 1, 26, 29, 30–31, 42, 51, 59, 64, 69, 70, 75
  Carmeli-Kaye's version of, 2
  spinorial version of, 2

Identities,
  Bianchi, 14, 15, 16, 17, 20, 21, 47, 64, 68,
Independent variables, 29
Indices,
  dyad, 11
  raising and lowering, 6
  spinor, 6
Inertial coordinates, 83–86
Inertial frames, 83
Integral,
  action, 30
Internal group, 78, 81–82, 88, 89
Invariance of the laws of physics, 3
Invariant,
  Lanczos, 29, 30, 31, 39, 42, 43
Invariants,
  as Lagrangians, 29, 42
  quadratic, 2, 29, 30, 31–33, 42
Isotopic spin, 7

Kaye, M., 2, 64, 69

Lagrange equation, 20, 21, 27, 38, 49, 50, 52
Lagrangian,
  Carmeli's, 2, 30, 33–34, 35, 42, 64, 66, 67, 68, 75
  Carmeli-Kaye's, 2, 26, 27, 64, 69, 75
  Einstein's, 50, 59
  gravitational, 5, 51
  Hilbert's, 1, 26, 29, 30–31, 42, 51, 59, 64, 69, 70, 75
  linear, 1, 29, 42
  Nissani's, 30, 35
  nonlinear, 1, 4
  Palatini's, 31
  quadratic, 1, 2, 32
  tensorial quadratic, 35–42, 43
Lagrangian density,
  electromagnetic, 19
  free-field gravitational, 19, 26
  gravitational-field, 21, 27
  matter, 27, 30, 52, 70

Subject Index

Lagrangian family, 34–35
Lagrangians,
   high-order theories, 32
   linear and quadratic, 29–30
   one-parameter family of, 42
   quadratic, 29–43
   spinorial family of, 37
Lanczos, C., 32
Lanczos invariant, 29, 30, 31, 39, 42, 43
Lanczos scalar density, 35
Lanczos variational method, 43
Landau, L.D., 53, 54, 55, 56, 62
Landau-Lifshitz complex, 53–56
Landau-Lifshitz pseudo-tensor, 53–56
Landau-Lifshitz super-potential, 56
Laws of physics,
   invariance of, 3
Leibniz rule, 8
Leibowitz, E., 2, 78
Levi-Civita symbol (metric), 6
Levi-Civita tensor, 25
Levi-Citiva tensor density, 14
Lifshitz, E.M., 53, 54, 55, 56, 62
Linear and quadratic Lagrangians, 29–30
Linear Lagrangian, 1, 29
Lorentz group, 4, 8, 78, 81, 82
   representations of, 5
Lorentz transformation, 4, 81

Mach principle, 78
Malin, S., 2, 58, 63, 64, 65, 75
Malin's conservation law, 63
Malin's current, 2, 64, 65, 66, 68, 75
Manifold,
   flat 4-dimensional, 5
   space-time, 4
Manifold mapping group (MMG), 3, 7, 8, 17
Mapping between tensors and spinors, 10
Matrices,
   Pauli, 10, 72
Matter,
   coupling with, 20
Matter energy, 48
Matter energy-momentum tensor, 51
Matter Lagrangian density, 27, 30, 52, 70
Maxwell equations, 19

Maxwell field tensor, 48
Metric,
   Levi-Civita, 6
   Tolman, 64, 71–74, 75
Metric equation, 21, 25
Metric structure of space-time, 9
Metric tensor, 6, 9, 29, 30, 31, 36
Minimal coupling, 45
Minkowskian coordinates, 9
Mixed spinors, 7
MMG (Manifold mapping group), 3, 7, 8, 17
MMG gauge theory, 4
Moller, C., 53, 57, 58, 59, 60, 62
Moller's approach, 56–60
Moller complex, 56–60
Moller pseudo-tensor, 56–60
Moller super-potential, 58

Newman-Penrose equations, 4, 23, 25, 34, 42
Newman-Penrose formalism, 2
   relation to the SL(2, C) gauge theory of gravitation, 21–25
Newtonian limit, 32
Nissani, N., 2, 30, 35, 46, 60, 62, 63, 69, 75, 78
Nissani conserved tensor density, 46
Nissani current, 4, 64, 69, 75
Nissani Lagrangian, 30, 35
Noether theorem, 2, 63, 64, 66, 67, 68, 70, 75
Non-Abelian gauge theory, 3
Nonlinear Lagrangian, 1, 4
Non-rotating frames, 77–89
Normalization condition, 7, 11, 21
Null tetrad, 22, 25, 26

One-parameter family of Lagrangians, 42
Ordinary divergence, 45, 46

Palatini, A., 25, 29, 31
Palatini action, 31
Palatini Lagrangian, 31
Palatini variational method, 4, 26, 43, 70
   derivation of SL(2, C) gauge field equations by, 25–27
Partial derivative,

dyad components of, 24
Particle physics, 1
Pauli matrices, 10, 72
Physics,
  particle, 1
  theoretical, 1
Poisson equation, 32
Potential,
  gauge, 4, 11–12
Potentials, 3
Preferred coordinate systems, 2, 78
Principle,
  action, 31, 42
  Mach, 78
  variational, 4, 19, 25, 30
Principle of equivalence, 4, 46
Principle of general covariance, 3, 45, 46
Product representation, 5
Proper Lorentz group, 4
Pseudo-tensor,
  Einstein, 49–53
  Landau-Lifshitz, 53–56
  Moller, 56–60

Quadratic invariants, 2, 29, 30, 31–33, 42
Quadratic Lagrangian, 1, 2, 32
  Carmeli, 2, 30, 30–34, 35, 42, 64, 66, 67, 68, 75
  Nissani, 30
  tensorial, 35–42
Quadratic Lagrangians, 29–43
  linear and, 29–30
Quadratic-Lagrangian formulation of general relativity, 2
Quantization program of the gravitational field, 42

Raising and lowering indices, 6
Relation of the SL(2, C) gauge theory to the Newman-Penrose formalism, 21–25
Relations between spinors and tensors, 9–11
Relativistic physics, 2
Relativity,
  general, 1, 4, 32
  special, 4, 45, 46, 48

Representations,
  double-valued, 4
  Lorentz group, 5
  product, 5
Ricci scalar, 18, 30, 47, 65
Ricci spinor,
  trace-free, 22
Ricci tensor, 18, 19, 47, 65
Riemann curvature tensor, 2, 3, 15, 16–19, 20, 29, 30, 31, 36, 42
  decomposition of, 18, 65
Riemannian space-time, 2
Rosen bi-metric theory, 58

Scalar,
  Ricci, 18, 30, 47, 65
Scalar densities, 32, 36, 37
Scalar density,
  Lanczos, 35
Scalar product, 7, 8
Scalars, 8, 11
Schwarzschild solution, 32
Set of six currents, 64-66
SL(2, C) conservation laws of gravitation, 63–75
SL(2, C) gauge theory,
  gravitation as an, 11–19
SL(2, C) gauge theory of gravitation, 1, 2, 3–37, 30, 46, 60
  Carmeli's quadratic Lagrangian for, 30
  conservation laws associated with, 2, 46
  fiber bundle formulation of, 2
  fundamentals of, 3–27
  relation to the Newman-Penrose formalism, 21–25
SL(2, C) group, 2, 4, 5, 7, 12, 42
SL(2, C) symmetry, 66–68
Source of the gravitational field, 21
Space,
  spinor, 2, 4, 11
  Tolman, 2, 71
Space-time,
  affine, 2
  curved 5, 10
  flat, 9
  metric structure of, 9
  Riemannian, 2

Subject Index

Space-time manifold, 4
Special relativity, 4, 45, 46, 48
Spin coefficients, 22
Spinor,
   contravariant, 5
   covariant, 5, 7
   dyad components of, 11, 12
   mixed, 7
   trace-free Ricci, 22
   valence of, 5
   Weyl, 22
Spinor affine connections, 8, 9, 10, 11, 12, 14
Spinor affine space-time, 2
Spinor algebra, 4–7
Spinor analysis, 4, 7–9
Spinor basis, 7, 11, 12, 21
Spinor calculus, 5
Spinor frame, 11
Spinor indices, 6
Spinor space, 2, 4–11
   conjugate, 5
Spinor-tensor,
   curvature, 14
Spinor transformation, 5, 6, 7, 8
Spinorial family of Lagrangians, 37
Spinors,
   correspondence between tensors and, 4
   relations between tensors and, 9–11
   two-component, 2, 5
Spinors at different points, 8
Superpotential, 56, 57, 62, 65, 68
SU(2) gauge field theory, 42
Symbol,
   Levi-Civita, 6
Symbols,
   Christoffel, 3, 30, 36, 47, 79
Symmetry and conservation laws, 63–64

Tensor,
   Carmeli's gauge field, 65
   conformal, 18, 19, 20
   curvature, 3, 20
   Einstein, 30, 47, 54
   electromagnetic stress-energy, 48
   energy-momentum, 2, 18, 31, 36, 38, 45, 48, 51, 78, 80, 87, 88
   gravitational, 36
   Levi-Civita, 25
   matter energy-momentum, 51
   Maxwell field, 48
   metric, 6, 9, 29, 30, 31, 36
   Ricci, 18, 19, 47, 65
   Riemann, 2, 3, 15, 16–19, 20, 29, 30, 31, 36, 42
   Weyl conformal, 18, 19, 20, 31, 43
Tensor calculus, 5
Tensor density,
   conserved skew-symmetric, 2
   Levi-Civita, 14
   Nissani's, 46
Tensorial quadratic Lagrangian, 35–42, 43
Tensors, 3
   correspondence between spinors and, 4
   relations between spinors and, 9–11
Tetrad,
   null, 22, 25, 26
Tetrad field, 58, 59
Theoretical physics, 1
Theory,
   Einstein's, 2
   field, 1
   gauge, 1
   Rosen bi-metric, 58
   SL(2, C) gauge, 1, 2, 3–27, 30
   Yang-Mills gauge, 8, 19
Theory of gravitation,
   Einstein's, 2, 4
   SL(2, C) gauge, 1, 2
Theory of Yang and Mills, 2, 4, 8
't Hooft, G., 42
Tolman metric, 64, 71–74, 75
Tolman space, 2, 71
Transformation,
   Lorentz, 4, 81
   spinor, 5, 6, 7, 8
   unimodular, 4
Trace-free Ricci spinor, 22
Transformation law for the gauge field, 13
Transformation law for the gauge potential, 12
Two-component spinors, 2, 5
Two-dimensional complex vector space, 4

Unimodular transformations, 4

Valence of a spinor, 5
Variation of $C_{\alpha\beta\gamma\delta}$ and $T_{\alpha\beta}$, 40–41
Variation of the metric tensor, 41–42
Variation of $\mathcal{L}$, 39–40
Variational principle, 4, 19, 25, 30
Vector space,
  two-dimensional complex, 4
von Freud, Ph., 56
von Freud super-potential, 56, 57

Weyl conformal tensor, 18, 19, 20, 31, 43
Weyl spinor, 22
Weyl-Eddington-type theories of gravitation, 32

Yang and Mills theory, 2, 4, 8
Yang-Mills current, 63, 75
Yang-Mills gauge theory, 8, 19